Exploring Mathematics with CAS Assistance

Topics in Geometry, Algebra, Univariate Calculus, and Probability

AMS / MAA | CLASSROOM RESOURCE MATERIALS

VOL 69

Exploring Mathematics with CAS Assistance

Topics in Geometry, Algebra, Univariate Calculus, and Probability

Lydia S. Novozhilova
Robert D. Dolan

An Imprint of the

MAA PRESS

AMERICAN MATHEMATICAL SOCIETY

Providence, Rhode Island

2020 *Mathematics Subject Classification.* Primary 97-01, 97D99, 97E99, 97F99, 97G99, 97H99, 97I99, 97K99, 97M99, 97N99.

For additional information and updates on this book, visit
www.ams.org/bookpages/clrm-69

Library of Congress Cataloging-in-Publication Data

Names: Novozhilova, Lydia S., 1948- author. | Dolan, Robert D. (Robert Domenic), 1993- author.
Title: Exploring mathematics with CAS assistance : topics in geometry, algebra, univariate calculus, and probability / Lydia S. Novozhilova, Robert D. Dolan.
Description: Providence, Rhode Island : MAA Press, an imprint of the American Mathematical Society, [2022] | Series: Classroom resource materials, 1557-5918 ; volume 69 | Includes bibliographical references and index.
Identifiers: LCCN 2022034189 | ISBN 9781470469887 (paperback) | ISBN 9781470472146 (ebook)
Subjects: LCSH: Mathematics–Data processing–Textbooks. | AMS: Mathematics education – Instructional exposition (textbooks, tutorial papers, etc.). | Mathematics education – Education and instruction in mathematics. | Mathematics education – Foundations of mathematics. | Mathematics education – Arithmetic, number theory. | Mathematics education – Geometry. | Mathematics education – Algebra. | Mathematics education – Analysis. | Mathematics education – Combinatorics, graph theory, probability theory, statistics. | Mathematics education – Mathematical modeling, applications of mathematics. | Mathematics education – Numerical mathematics.
Classification: LCC QA76.95 .N68 2022 | DDC 510.285/536–dc23/eng/20220825
LC record available at https://lccn.loc.gov/2022034189

To Alice and Beatrice

Lydia Novozhilova

To Desirée

Bobby Dolan

Contents

Introduction   xi

Acknowledgments   xv

Part 1   Algebra & Geometry   1

1   Computer Algebra Systems and Elements of Algorithmics   3
    1.1   Terms and notation   4
    1.2   About data types and data structures   5
    1.3   Elements of algorithmics and algorithmic problem solving   8
    1.4   Glossary   16

2   Topics in Classical Geometry   19
    2.1   Review: Matrices, vectors, and lines   19
    2.2   Rigid transformations of the plane   25
    2.3   Complex numbers in classical geometry   27
    2.4   Three centers of a triangle   30
    2.5   Glossary   33

3   More Topics in Classical Geometry   35
    3.1   Lab 2: The Euler line   35
    3.2   The Simson line   37
    3.3   Conics   40
    3.4   Glossary   45

4   Topics in Elementary Number Theory   47
    4.1   Number of primes and the Riemann Hypothesis   49
    4.2   Algorithms from elementary number theory   50
    4.3   Pythagorean triples   56
    4.4   Lab 4: Plotting legs of primitive Pythagorean triples   58
    4.5   Linear Diophantine equations in two variables   59
    4.6   Lab 5: Industrial application of an LDE in three variables   62
    4.7   Glossary   63

5   Topics in Algebra: Solving Univariate Algebraic Equations   65
    5.1   Roots of univariate polynomials   66
    5.2   Geometry of cubic equations: Counting the number of real roots   69
    5.3   Lab 6: Solving cubic equations using Vieta's substitution   74
    5.4   Nonnegative univariate polynomials   76

    5.5   Glossary   78

6   Topics in Algebra: Bivariate Systems of Polynomial Equations   79
    6.1   Linear systems of two equations   80
    6.2   Nonlinear systems of polynomial equations: Motivating example   82
    6.3   Solving nonlinear polynomial systems   85
    6.4   Implicitization of plane curves   94
    6.5   Glossary   96

Part 2 Calculus and Numerics   97

7   Derivatives   99
    7.1   Review: Definitions, notation, and terminology   100
    7.2   Convexity of a univariate function   104
    7.3   Some facts about functions and derivatives   105
    7.4   Lab 8: Constructing a square circumscribed about ellipse   110
    7.5   Glossary   112

8   Definite Integrals   113
    8.1   Review: Some basic concepts and facts of univariate integral calculus   114
    8.2   Area of a region bounded by a simple closed curve   116
    8.3   Lab 9: Submergence depth of a body of revolution in equilibrium   121
    8.4   Solving some ordinary differential equations   123
    8.5   Glossary   127

9   Approximating Zeros of Functions by Iteration Methods   129
    9.1   Fixed point iteration method   130
    9.2   Newton's method   135
    9.3   Lab 11: Kepler's Equation and deriving Kepler's Second Law   137
    9.4   Lab 12: Exploration of the logistic maps   140
    9.5   Glossary   141

10 Polynomial Approximations   143
    10.1 Taylor polynomials   145
    10.2 Interpolating polynomials in the Lagrange form   147
    10.3 Piecewise polynomial interpolation: Splines   150
    10.4 Approximating large data sets: Regression   153
    10.5 Two real-life applications of the LS method   156
    10.6 Glossary   158

11 Trigonometric Approximation   159
    11.1 Short review of trigonometric functions   160
    11.2 Fourier series   164
    11.3 About the accuracy of trigonometric approximations   167
    11.4 Celebrated classical application of Fourier series   170
    11.5 Glossary   173

12 Fourier Analysis in Music and Signal Processing   175
    12.1 Introduction and background   175

12.2 Fourier series and periodic signals 177
12.3 The Fourier transform for non-periodic signals 180
12.4 The Discrete Fourier Transform 182
12.5 Fourier series in signal processing 187
12.6 Glossary 188

Part 3 Probability and Statistics 191

13 Probability and Statistics Basics 193
13.1 Review: Some basic concepts of probability 194
13.2 Some discrete probability distributions 197
13.3 About continuous probability distributions 201
13.4 Law of Large Numbers 204
13.5 Central Limit Theorem 206
13.6 Glossary 208

14 Computer Simulation of Statistical Sampling 209
14.1 Random number generation 209
14.2 Lab 17: CLT and LLN in action: Life expectancy in the world
population 211
14.3 Sampling from non-uniform distributions (optional) 213
14.4 Monte Carlo methods for finding integrals and areas 215
14.5 Lab 18: Buffon's needle problem 222
14.6 Glossary 223

15 Simple Random Walks 225
15.1 Simple random walks on integers 226
15.2 Lab 19: The gambler's ruin problem 231
15.3 Random walk on the square lattice 233
15.4 Lab 20: Drunken sailor problem 234
15.5 Glossary 235

A Data for Lab 17 in Chapter 14 237

Bibliography 239

Index 241

Introduction

This book originated from lecture notes for different versions of a one-semester course called "Symbolic Computations" that the first author taught for ten years at Western Connecticut State University. It was a hands-on, problem-solving, and project-oriented course that included topics from number theory, classical geometry, algebra, univariate calculus, and probability. The selected topics and problem-solving techniques from these fields were illustrated through various computational examples and visualizations, and most exercises, mini projects, and labs were intended to be completed using computer assistance.

In worked out examples, students observed and, in their own solutions, demonstrated how the power of mathematical tools can be greatly enhanced by the power of computational software. For seven years the course was taught using the computer algebra system (CAS) Maple. We then switched to SageMath[1], a CAS written in the popular programming language Python. However, the rapid development and dynamic change of the software used in the course required continuous time-consuming updates of the instructional material. For the same reason, many nice books with content that could be used in the course, such as [16], quickly became outdated. This is the reason for choosing a software-independent presentation in this text. The choice of specific software is left to the instructor or reader.

The focus of our presentation is on mathematical tools and their uses in problem solving, not on the technicalities of any specific CAS. We emphasize the mathematical underpinning of solution methods and their **algorithmic structure**. While the term **algorithm** is ubiquitous in computer science, it is not quite as common in mathematical educational literature (although any method in computational mathematics presented as a step-by step sequence of verbal instructions is a representation of an algorithm). We believe that recasting such verbal instructions into a crisp and succinct algorithmic language is beneficial for mathematics students and enhances their understanding of the connections between mathematics and computer science.

Computer assistance in solving problems in this book is intended to be used for:

- strengthening students' understanding of the logical structure of mathematical methods and their algorithmic implementations

- handing over tedious and repetitive mathematical calculations to computers

- creating beautiful and meaningful mathematical displays

---

[1] A very comprehensive presentation of using SageMath for mathematics can be found in [12].

There are 111 figures in this text illustrating mathematical concepts, statements, and solutions to examples. Creating similar mathematical displays with computer assistance is an important part of the overwhelming majority of this book's tasks. Without a doubt, visualizing one's solutions provides a deeper understanding of the problem at hand.

Prerequisites needed for working with this text are rather modest. Formally, a working knowledge of discrete mathematics and univariate differential and integral calculus is sufficient. We do provide brief reviews of topics and concepts that are essential for solving included problems, such as the median of a triangle, conics, prime numbers, and derivatives. Certain terms that might not be familiar, for example, operations on small-sized (2×2) matrices, are carefully introduced and demonstrated with examples. We do not intend in this book to teach concepts; we want to show some concepts and methods in action and involve students in doing the same by solving various and often nonstandard problems. Programming experience and some basic familiarity with elementary probability and statistics are beneficial but not required. If desired, help with code writing can be provided to students in many different ways inside and outside of the classroom. For example, instructors can provide brief tutorials, code templates, workshops, and other resources to help students get started.

This book includes fifteen chapters that are split into three parts (modules). Part 1, Geometry and Algebra, and Part 2, Calculus and Numerics, include six chapters each; Part 3, Probability and Statistics, includes three chapters – on elementary probability, computer simulation of random numbers, and random walks. Chapter 1 in Part 1 is very important for the computer programming components of this text. Here we briefly introduce some basic CS terminology for data types and two basic control flow statements (loops and conditionals). In addition, we include examples of several simple algorithms using pseudocode. Furnished with codes in any general-purpose CAS, these pseudocodes, as well as code for any computational exercise in this book, are short procedures (often called **functions** in computer science) that are typically only about ten lines of commands.

The text includes more than one hundred numbered exercises, a large number of worked-out examples, sixteen mini projects, and twenty labs. Mini projects and labs often involve exploratory-type questions and making appropriate CAS functions. Examples illustrate important concepts, techniques, and their applications, and provide verbal descriptions of the algorithms used. Mini projects and labs are project-type assignments that require more substantial time and guidance. Some of the labs may be (and have been) used as starting points for midterm projects, semester projects, and even presentations at undergraduate and professional conferences. In addition, there are numerous simple, short tasks that do not require computer assistance. Questions like "Why?" and phrases like "Show this!", as well as unnumbered exercises marked by the sign ⧖, pepper the text and are intended to test students' understanding of the text fragment directly preceding the question.

A remarkable mathematician and educator P. Halmos saw the work of a mathematician as "…similar to what a laboratory technician does"[1]. Inspired by this view, we wanted the book to be a constructive and hands-on problem-solving source that provides appropriate guidance, but with some space for "trial and error, experimentation, and guesswork"[1]. This is one reason why we strived to create original, nonstandard assignments. In the same token, when writing directions for a multi-step problem, we used the phrase "**Suggested** directions" to give motivated students an option to design their own path for solving the problem. In numerous exercises and in about half of mini projects and labs the readers will be dealing with applications of mathematical techniques to problems in science, and with this first-hand experience they can appreciate "the unreasonable effectiveness of mathematics in the natural sciences." (This is the title of a paper[2] by Eugene Wigner, a Nobel prize winner in physics.)

For an easy version of a course based on this text we suggest selected materials from Chapters 2 through 8. For a challenging course selected materials from Chapters 9 through 15 can be used, plus preliminary independent reading and short in-class reviews of the relevant parts of Chapters 7 and 8. Chapter 1 should always be included.

In addition to being used as a text for a dedicated course, this book can also be used as a textbook for a problem-solving or topics course. The modular structure of this text allows the contents to be used selectively, rearranged, and augmented at the instructor's discretion. This feature provides instructors the opportunity to naturally tailor the course towards a specific student audience. Classes with both math and CS majors are especially effective due to the possibility of student interaction and teamwork on projects and labs. Other possible uses of the book could be the following:

- As a source for worked out examples and nonstandard assignments that can be used to enliven various mathematical courses related to the fields the book touches upon.

- As a supplementary source in a topics course with a "do-it-yourself" computer component.

- As a source of topics and projects for seminars and presentations.

- For self-study by those who want to refresh their memory of some basic college-level mathematics and enhance it by using computer assistance for solving constructive and exploratory computational problems.

Many remarkable mathematicians, physicists, and computer scientists, as well as education policy makers, have emphasized how important the interaction between computer science, mathematics, and their applications is. There are excellent textbooks that present certain methods and tools from these three fields in a constructive and coherent manner. We hope that our book is a modest contribution in this direction. While working on it, we again and again felt the special beauty, utility, and unity of the world of mathematics, even at the elementary level of this text. We would be happy if, in addition to enhanced proficiency in solving mathematical problems with computer assistance, readers will also experience this feeling.

---

[2]Communications in Pure and Applied Mathematics, Vol. 13, issue 1, 1-14.

Acknowledgments

A precursor of this book based on different versions of a course that I taught for a number of years was written during my sabbatical leave in the fall semester of 2018. I am very grateful to the Academic Leave Committee at Western Connecticut State University for approving my sabbatical proposal.

Based on this work, I wrote a manuscript and a draft was sent to Robert Kennedy in 2019. He detected some good ideas in it and showed the text to Cynthia Hoffman; I am deeply indebted to both of them. Cynthia, in addition to constructive comments and critiques, suggested to find a native speaking coauthor. It was another stroke of luck that my former student, Robert Dolan, who is now an Assistant Professor at University of Connecticut, agreed to play this role. The first draft has been substantially reworked and expanded upon after Robert and I decided to write this text. Without Robert's numerous and thorough corrections of the original text, his invaluable expertise in Latex, and our constructive (and also enjoyable) discussions of the contents of the book, it would not have come to life. In addition, Robert added a chapter on Fourier analysis in music (and he has expert knowledge in both).

Inspiration for developing this text came from various sources including books that I read as a student and later from numerous documents found on the web written by pedagogically and mathematically talented people. The most prominent are the books [10], [13], [11], and [21]. The idea to tie together content from various mathematical fields and computational recipes was inspired by the books [8] and [9]. Last but not least, a great source of inspiration was students who took my course and expressed their appreciation of its outcomes even years after completing it.

Lydia Novozhilova, July 2022

Our great thanks go to the anonymous reviewers of the CRM Editorial Board who helped tremendously in improving and expanding our manuscript. The text would not be what it is today without their contributions.

Part 1

Algebra & Geometry

1

Computer Algebra Systems and Elements of Algorithmics

Computer Algebra, also called Symbolic Computation, studies the development of algorithms and software for operations on symbols rather than numbers. Symbols represent mathematical objects, such as numbers, expressions, functions, equations, and other mathematical structures. Professional algorithm designers strive to design algorithms that implement a particular task in the most efficient way possible. Algorithmic efficiency is a property that is related to the amount of computational resources used by an algorithm. Efficiency of an algorithm is formalized by its complexity. Computational complexity of an algorithm is a measure of the amount of computational resources – time and space – that an algorithm consumes when it runs. Although computational complexity is an important property when developing and analyzing computational algorithms, we will not discuss it in detail here since the focus of this text is on certain basic problem solving methods and the underlying mathematics of implementing these methods on computers. The amount of resources needed for computational tasks in this book is very modest.

We assume that an existing computer algebra system will be used to assist in solving problems that involve tedious symbolic transformations or numerical computations and plotting. The term "computer algebra system" (CAS) refers to a mathematical software system that is capable of performing both symbolic and numerical computations. The core functionality of a CAS is to exactly manipulate symbolic mathematical expressions. In addition to standard arithmetic operations, a general purpose CAS typically includes

- Polynomial manipulations, such as substitution, expansion, factorization, and division

- Calculus operations, such as finding limits, differentiation, indefinite and definite integration, and constructing Taylor series

- Finding exact solutions of various types of equations, such as polynomial equations, differential equations, and systems of linear equations

- Geometric manipulations

- Plotting capabilities for various kinds of mathematical displays

More advanced CASs include sophisticated functions and packages designed for specific scientific fields and applications, but we will not discuss such features here. Presently, Maple and Mathematica are considered as the major general purpose proprietary CASs. Open-source CASs have also been developed, such as SageMath and SymPy.[1] Some CAS-equipped calculators are also available.

1.1 Terms and notation

Collections of elements. A **set** is an *unordered* collection of *distinct* (non-repeated) elements. A finite set can be described by listing all of its elements, for example, $S = \{1, 2, 3, 4\}$. In an abstract mathematical setting, formal set-builder notation $S = \{a \mid \dots\}$ (or $S = \{a : \dots\}$) is used, where the ellipsis "..." stands for a characteristic property shared by all elements of the set. The vertical bar reads "such that". For example, $\{x \mid 0 \le x \le 1\}$ defines the set of all points in the closed interval $[0, 1]$. In this text we will use the notation $S = \{a_0, a_1, \dots, a_{n-1}\}$ or $S = \{a_j\}_{j=0}^{n-1}$ for a finite set S with n elements whose characteristic property is clear from the context. An infinite set with a clear characteristic property of its members is denoted as $S = \{a_0, a_1, \dots\}$.

A **sequence** is an *enumerated* collection of objects in which repetitions are allowed and order matters. In mathematical settings in this text, a finite sequence S with n elements is denoted as $S = (a_0, a_1, \dots, a_{n-1})$ or $S = (a_j)_{j=0}^{n-1}$. Analogous to an infinite set, an infinite sequence is denoted as $S = (a_0, a_1, \dots)$. For example, $S = (2, 4, 6, 8, \dots)$ is an infinite sequence of even nonnegative integers. In computer science (CS), a sequence is considered as a data structure that has three "incarnations": as a list, as a string, and as a tuple. In the next section, we will introduce these data types along with a few basic operations available for each of them in most general purpose programming systems.

Notation for standard sets of numbers.

- $\mathbb{N} = \{1, 2, 3, \dots\}$ – the set of natural numbers

- $\mathbb{N}^0 = \{0, 1, 2, 3, \dots\}$ – the set of nonnegative integers

- $\mathbb{Z} = \{\dots, -2, -1, 0, 1, 2, \dots\}$ – the set of integers

- $\mathbb{Q} = \{m/n : m, n \in Z\}$ – the set of rational numbers

- $\mathbb{R} = \{x : x \in (-\infty, \infty)\}$ – the set of real numbers

- $\mathbb{C} = \{x + iy : x, y \in R, i^2 = -1\}$ – the set of complex numbers

- The notation $\mathbb{R}^2$ stands for the set of ordered pairs (x, y) of real numbers or, equivalently, the set of points on the plane with rectangular coordinates x and y.

- If a is an element of a set or list M, we write $a \in M$.

---

[1]SageMath is a free, open-source CAS licensed under the GPL. It uses syntax resembling Python, a very popular and broadly used programming language in academia. SymPy is a free software system written in Python and licensed under the BSD. It aims to become a full-featured CAS.

Other notation.

- Two expressions with the sign "$\equiv$" between them introduce a new mathematical object on the left-hand side defined by the right-hand side. For example, $p \equiv x^2 + 2x + 5$ gives the name p to the quadratic expression on the right-hand side, which is a convenient shortcut for future use of this expression.

- The notation $\lfloor x \rfloor$ and $\lceil x \rceil$ stand for the value of the floor and ceiling functions, respectively, at x. The floor function $\lfloor x \rfloor$ is defined as the largest integer not exceeding x, and the ceiling function $\lceil x \rceil$ is the smallest integer greater than or equal to x. For example, $\lfloor 3.48 \rfloor = 3$, $\lfloor -1.15 \rfloor = -2$, $\lceil -0.57 \rceil = 0$, and $\lceil 3 \rceil = 3$.

- Let $(a_j)_{j=0}^n$ be a sequence of numbers or expressions. The notation $\sum_{j=0}^n a_j$ stands for the sum $a_0 + a_1 + \cdots + a_n$. The product of the elements of this sequence is denoted as $\prod_{k=0}^n a_k$.

- A **neighborhood** of a point $x \in \mathbb{R}$ is any interval (a, b) that contains this point and we write $x \in (a, b)$.

- The notation $\lim_{x \to a^{\pm}} f(x)$ stands for one-sided limits of a function f at a point $x = a$. Corresponding values of the one-sided limits are denoted as $f(a^{\pm})$.

- If the truth of a statement p implies that of q, we write $p \Rightarrow q$ (pronounced "if p then q"). For example, if p is the statement "The series $a_1 + a_2 + a_3 + \ldots + a_n + \ldots$ converges", and q is the statement $\lim_{n \to \infty} a_n = 0$, then $p \Rightarrow q$. Note that if the statement $p \Rightarrow q$ is true and q is false, then p is also false.

- If the truth of a statement p is equivalent to that of q, which means that they are either both true or both false, we write $p \Leftrightarrow q$ (pronounced "q if and only if p"). For example, suppose a function f is defined in a neighborhood of a point $x = a$. If p is the statement $\lim_{x \to a^-} f(x) = \lim_{x \to a^+} f(x) = c$ and q is the statement $\lim_{x \to a} f(x) = c$, then $p \Leftrightarrow q$.

Text symbols.

- The sign $\boxed{\text{\small ☼}}$ marks short simple exercises. Their purpose is to check the reader's understanding of a small portion of material immediately prior to this sign.

- The sign $\boxed{\text{\small 📖}}$ is used for either historical remarks or comments on something beyond the scope of the text.

1.2 About data types and data structures

There are numerous sources for a concise introduction to computer programming. In this section we just briefly and informally introduce some basic terms and data structures in computer programming that will be useful for the reader in completing computer assisted assignments.

A **data type** is a classification of data which tells the compiler or interpreter what values and operations are available for this type.[2] Programming languages support

---

[2]Data types can be declared in different programming languages using different syntax.

primitive and **composite** data types. Primitive data types, such as real, integer, complex, Boolean, and character, are the most fundamental data types usable in a programming language. Primitive data types are used to construct composite data types, such as set, string, list, array, and dictionary.

A **string** is a data type representing a sequence of alphanumeric text or other symbols. It is customary in computer science to use single or double quotes for strings. For example, below are strings named *greet*, *nonsense*, and *pandemic*:

```
greet="Hello world!"
nonsense="Abracadabra"
pandemic="Covid-19"
```

We can merge two strings into a new one or split a string into substrings and construct a list of these substrings. (The latter operation is used in the last example of this chapter.) Various operations on strings are widely used in text processing.

Lists are one of the most powerful tools in computer science and will be used in numerous computational assignments in this text. A list L with n elements is denoted as $L = [a_j, j = 0...n - 1]$ or $L = [a_j]_{j=0}^{n-1}$. List indexing in this text will typically start from zero. For example, given the list $L = [-1, 5, 5]$ of length three, we have $L[0] = -1, L[1] = 5$, and $L[2] = 5$. Lists are mutable, i.e., they can be altered once declared. In particular, one can add an element to a list, remove an element from a list, or replace an element with a different one. Lists in programming can include different types of elements. For example, the list

```
G=["Alice",8,"Beatrice",3]
```

contains two strings ($G[0]$ and $G[2]$) and two integers ($G[1]$ and $G[3]$).

Tuples provide a mechanism for grouping and organizing data to make it easier to use. A tuple lets us "chunk" together related information and use it as a single object. Here are two examples of tuples:

Example 1.1.

```
point=(x,y,z)
Roberts = ("Julia", 1967, "Atlanta", "Georgia", "Actress")
```

Tuples and strings are considered as immutable, or unchangeable, data types in many computer languages. This means that elements cannot be added to or removed from a tuple or string. However, existing tuples and strings can be used to create new tuples or strings. The meaning of individual elements of a tuple can usually be understood from the presented context. For example, one can infer that in the tuple above named *Roberts*, "Julia" is the first name, 1967 is the year of birth, and the last two elements are the place of birth and profession, respectively. But this can be made more clear if we use another data structure called a dictionary.

A **dictionary**, also called an associative array, map, or symbol table, is a data type composed of a collection of (*key, value*)-pairs, whose role in presenting information in the dictionary will be clear from the example below. Each possible key appears at most once in the collection. Here is an example of a dictionary related to statistics called a *five-number summary* (in Python syntax):

```
d={"min": 1.3, "Q1": 2.1, "median": 4.0, "Q2": 5.7, "max": 5.2}
```

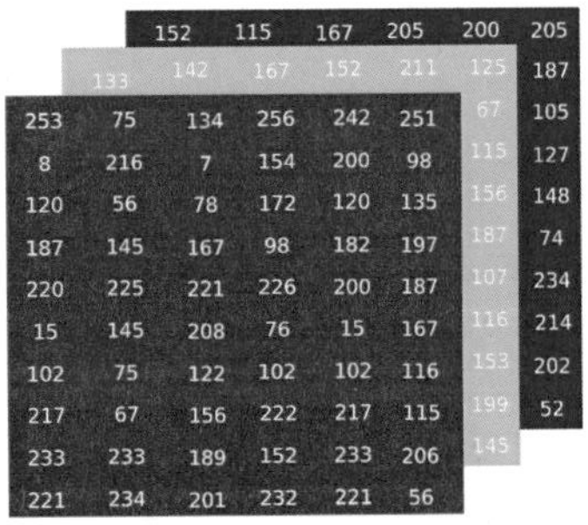

Figure 1.1. A visual example of the RBG model.

In this dictionary all of the *keys* are strings and the number following each key is the corresponding *value*.

Exercise 1.2. Convert the tuple named *Roberts* in Example 1.1 into a dictionary using the syntax shown in the dictionary example above.

Remark 1.3. Programming languages provide certain operations (functions) for each specific data type. For example, operations on lists that will be used often in this text typically include replacing/removing an element of a list or appending a new element.

A **data structure** is a collection of data type values which are stored, organized, and managed in a format that allows for efficient access and modification. In turn, all composite data types are data structures. For example, an **array** is a data structure consisting of a collection of elements, each identified by at least one array index. Analogous to the mathematical concepts of a vector and matrix, array types with one and two indices are often called **vector type** and **matrix type**, respectively. More generally, a multidimensional array type can be called a **tensor type**.

Example 1.4. The digital representation of a grey image is a two-dimensional array (matrix). A colored image can be represented using, for example, the RGB model, by an $M \times N \times 3$ array. In this array, M is the y-dimension, N is the x-dimension, and the third dimension is the "color dimension," which represents the colors red, green, and blue. The RGB model is a so called additive model in which red, green, and blue colors are added together in various ways to reproduce a broad spectrum of colors. In other words, each element of the $M \times N$ matrix is a vector of length three which represents the "RBG value". For example, the RGB value $(128, 0, 128)$ defines the color purple of a pixel at the location of this value in the $M \times N$ array. See Fig. 1.1 for a visual interpretation.

A very general term in computer science is an **object**. It can be a variable, data structure, function, or method, and as such, is a value in memory referenced by an **identifier**. An identifier is a name that is assigned to an object by the user. It is usually limited to letters, digits, and underscores. For example, one could define a function called *disk_area* that takes the radius r of a disk and returns the area $\pi \cdot r^2$. Here, *disk_area* is the identifier of the function. This book includes numerous exercises that involve making functions to execute specific mathematical tasks.

1.3 Elements of algorithmics and algorithmic problem solving

Algorithmic problem solving is a course in its own right taught in most CS departments. This section includes just some cursory comments and gives a few examples of simple algorithms and their descriptions. An algorithm can be informally described as a step-by-step procedure by which an operation can be carried out without any exercise of intelligence, and so, for example, by a machine. Formally, an algorithm is a specification of a procedure by which a given type of problem can be solved in a finite number of mechanical steps. Simple algorithms familiar from elementary arithmetic are finding the greatest common divisor of two integers and synthetic division of two polynomials.

1.3.1 Control flow statements. It is common when solving a mathematical problem that one has to make a choice which of two or more computational paths to follow. Often this choice depends on a certain intermediate result. For example, if the task is finding real roots of a quadratic equation, and one calculated the discriminant D, the next step could be calculating the real roots if $D \geq 0$ or, otherwise, making a statement "The equation does not have real roots."

Encoding such situations on a computer involves using so called **conditional statements**, or simply **conditionals**. Conditional statements tell the computer to execute specific actions *provided certain conditions are met*. Loops and conditional statements are examples of control flow mechanisms that are routinely used in programming. Below is a simple function using an *if* conditional statement:[3]

```
def is_positive(x):
    if x>0:
        print("x is positive")
```

When called with a specific input x, the function simply verifies whether the condition $x > 0$ is true or false. If the condition is false, the function returns nothing.

When there are two choices for a possible computational path, the conditional *if-else* can be used; for three possible options to continue computations, a computational system typically provides some version of the *if-elif-else* conditional. Here is a simple example:

```
def test_sign(x):
    if x > 0:
        print("x is positive")
    elif x < 0:
        print("x is negative")
    else:
        print("x=0")
```

Generally, the block of code for each option can include more than one command. Sometimes it is convenient to use what is called a *compound conditional*. Compound conditionals are a way to test two conditions in just one statement. Here is an example:

---

[3]We are using Python syntax in the next three examples.

```
def is_zero(x):
    if (x > 0) or (x<0):
        print("x is a nonzero number")
    else:
        print("x=0")
```

Some calculations, like finding the sum of elements of a numerical sequence, involve repeated mathematical operations. Implementing such repeated calculations on a computer uses another type of control flow statement called a **loop**. A loop is a programming structure that repeats a sequence of instructions until a specific condition is met. Many of the computational assignments in this book require the reader to use *for* and *while* loops. A *for* loop is a block of code that is repeated a fixed number of times. A *while* loop continuously executes the body of the loop so long as some condition remains true. The condition appears to the right of the keyword *while*. Here are simple, toy examples demonstrating the ideas of the two types of loops.

Example 1.5.

(a) Suppose a robot located at the origin can move along the x-axis only by steps of length two, and the task is to reach an object located at the point $x = 20$. Set the variable x, position of the robot, to zero, and let j be a **loop variable**, the variable for counting steps. Then a *for* loop for this task can be presented (not in any specific language) as

```
for j from 0 to 9
    x=x+2
```

The counter j starts with zero value and after each update of position (adding 2 to the current x-value) is increased by one. "Behind the scene", computer checks the counter, and if $j < 10$, repeats the command $x = x + 2$.

(b) The same task can be implemented using a *while* loop as follows. (Again, initially x is set to zero.)

```
while x<20:
    x=x+2
```

Notice that there is no a loop variable in this *while* loop, and the number of iterations is defined by the task itself.

✐ *Check Your Understanding.* Pretend that you are a computer and manually implement computations described in this example. What is the value of the loop variable at the end of the last iteration in the *for* loop? What is the value of the position variable at the beginning of the last iteration in the *while* loop?

The reader finds examples of using *for* and *while* loops later in this section.

1.3.2 Writing algorithms. An algorithm can be expressed in various ways, including natural language, flowcharts, and programming language. Directions for getting from one place to another or cooking instructions are everyday examples of algorithms written in natural language. Algorithms described in this way are easier for a wide audience to understand, but are often inefficient and vague. To overcome these issues with using natural language, programmers commonly express algorithms in what is called **pseudocode**. Pseudocode employs the structure of formal programming languages and mathematics to split the algorithm into one-sentence steps. No broad standard for pseudocode syntax exists and exactly how to structure pseudocode is a personal choice. Below we demonstrate this crisp and concise way of presenting simple mathematical methods in computer science. Note that writing and understanding pseudocode is not required for this text, but is an optional tool that will undoubtedly help when approaching the CAS exercises, projects, and labs.

The main goals in writing an algorithm should be clarity and precision. Here are some commonly accepted rules:

- Use control flow statements, such as **if-else, if, while, for**, and **return**, to control the order in which individual statements, instructions, and function calls are executed.

- Use standard imperative programming keywords and notation, such as *variable ← value, array[index], function(args)*.

- All groups of statements showing dependency are to be indented. These include **while, do, for, if**. Indent everything carefully and consistently.

- Don't typeset keywords in a different font or style.

- Each statement should fit on one line, and each line should contain only one statement. (The only exception is extremely short and similar statements like $i = i + 1; k = 0$.)

- Put each structuring statement (**for, while, if**) on its own line. The order of nested loops matters a great deal; make it absolutely obvious.

- Use short, but mnemonic, algorithm and variable names.

A well-written algorithm description can be implemented easily by any competent programmer in any programming language, even if they don't understand why the algorithm works. Good pseudocode, like good code, makes an algorithm easier to understand and analyze, while also making mistakes easier to spot. For a systematic presentation of algorithmics, see, for example, the very readable book [**4**].

1.3.3 Pseudocode examples.

Example 1.6. Let the elements of an array be stored in a list T of length n. The elements can be real numbers or mathematical expressions. The following is pseudocode for finding the sum of all elements in the array:

Algorithm Finding array sum

1:	**function** ARRAYSUM(T)	▷ The sum of elements in list T
2:	$s \leftarrow 0$	▷ Set s to 0
3:	**for** $i \leftarrow 0, n - 1$ **do**	▷ Loop variable varies from 0 to $n - 1$
4:	$s \leftarrow s + T[i]$	▷ Update s
		▷ When $i = n$, the loop is complete
5:	**return** s	▷ The sum is s

Example 1.7. Manually implement the algorithm in Example 1.6 to find the sum of the elements in the list $[3, -1, 8]$.

Solution. Note that here $n = 3$.

(1) $T = [3, -1, 8]$

(2) $s = 0$

(3) $i = 0$

(4) $s = s + T[0] = 3$

(5) $i = 1$

(6) $s = s + T[1] = 2$

(7) $i = 2$

(8) $s = s + T[2] = 10$.

(9) $i = 3$. The loop is complete.

Answer: The sum of the elements in T is 10.

Example 1.8. Let the elements of an array be stored in a list T of length n. It is assumed here that the elements are real numbers. The following is pseudocode for finding the minimal element in the array:

Algorithm Finding minimal element

```
1: function MINVALUE(T)                          ▷ The minimal element of T
2:     min_val ← T[0]                                   ▷ Set min_val to T[0]
3:     for i ← 1, n − 1 do            ▷ Loop variable varies from 1 to n − 1
4:         if T[i] < min_val then
5:             min_val ← T[i]              ▷ Update min_val if condition holds
                                           ▷ When i = n, the loop is complete
6:     return min_val
```

Example 1.9. Manually implement the algorithm in Example 1.8 to find the minimal element in the list $[3, -2, 0]$. (Note that $n = 3$.)

Solution.

(1) $T = [3, -2, 0]$

(2) $min\_val = T[0]$

(3) $i = 1$

(4) $T[1] = -2 < 3$

(5) $min\_val = -2$

(6) $i = 2$

(7) $T[2] = 0 > -2$.

(8) $i = 3$. The loop is complete.

Answer: The minimal element is $min\_val = -2$.

Example 1.10. Horner's method is an efficient polynomial evaluation method that does not require exponentiation. For a low degree polynomial the method can be implemented in one short formula. For example, we can find $p(2)$ for the quadratic polynomial $p(x) = 5x^2 - 3x + 1$ by just calculating $(5 \cdot 2 - 3) \cdot 2 + 1$. If one wants to repeatedly evaluate higher degree polynomials, it is reasonable to use computer assistance and make a function that implements Horner's method.

Let $p(x) = \sum_{j=0}^{n} a_j x^j$ be a polynomial of degree n. Define the list of coefficients $A = [a_n, a_{n-1}, \ldots, a_0]$. Notice that the length of the list A is $n+1$ and $A[0]$ is the leading coefficient a_n of the polynomial. Horner's method for evaluating the polynomial p can then be implemented via the following algorithm:

Algorithm Horner's method

1: **function** POLYVAL(A, c) $\triangleright$ Value of polynomial with coefficients A at $x = c$

2:     $val \leftarrow c$ $\triangleright$ Set val to c

3:     $result \leftarrow A[0]$ $\triangleright$ Set $result$ to the leading coefficient

4:     **for** $i \leftarrow 1, n$ **do** $\triangleright$ Loop variable varies from 1 to n

5:         $result \leftarrow result \cdot val + A[i]$ $\triangleright$ Update $result$

                                       $\triangleright$ When $i = n + 1$, the loop is complete

6:     **return** $result$

Example 1.11. Manually implement the algorithm in Example 1.10 to evaluate the polynomial $2x^3 - 7x + 5$ at $x = 3.5$.

Solution.

(1) $A = [2, 0, -7, 5]$

(2) $val = 3.5$

(3) $result = A[0] = 2$

(4) $i = 1$

(5) $result = result \cdot val + T[1] = 7$

(6) $i = 2$

(7) $result = result \cdot val + T[2] = 17.5$

(8) $i = 3$

(9) $result = result \cdot val + T[3] = 66.25$

(10) $i = 4$. The loop is complete.

Answer: $p(3.5) = 66.25$.

Euclidean division, also called **division with remainder**, is a well known procedure of dividing one integer, called the **dividend**, by another integer, called the **divisor**. The procedure is used for different mathematical tasks, in particular, for finding the greatest common divisor of two positive integers m and n, commonly denoted by $\gcd(m, n)$.[4] The **Euclidean algorithm** is an efficient method for finding $\gcd(m, n)$ by repeatedly using Euclidean division. The next example shows an informal manual implementation of the algorithm.

Example 1.12. Find $\gcd(18564, 63)$.

Solution.

(1) Implementing Euclidean division, we have $18564 = 63 \cdot 294 + 42$. Here 18564 is the dividend, 63 the divisor, and 42 the **remainder**. It follows that the common divisor of 18564 and 63 is also the common divisor of 63 and 42. (**Think why.**)

(2) Choosing 63 as the next dividend and 42 as the next divisor, we obtain $63 = 42 \cdot 1 + 21$. Here 21 is the remainder of dividing 63 by 42.

(3) Now setting the new dividend to 42 and the new divisor to 21, we obtain $42 = 21 \cdot 2 + 0$. The remainder is zero. Thus, $\gcd(18564, 63) = 21$.

Below we provide pseudocode that presents the pattern shown in the above example more formally. The pseudocode includes the modulo function (or modulo operation) denoted by $m \bmod n$. The modulo operation takes a pair of positive integers and returns the remainder of dividing the first integer by the second. It is a built-in function in any general purpose software.

Algorithm Euclidean Algorithm

1: **function** EUCLID(a, b)   ▷ The gcd of a and b
2:      $r \leftarrow a \bmod b$ ▷ Set r to $a \bmod b$
3:      **while** $r \neq 0$ **do**   ▷ Repeat until condition is false
4:          $a \leftarrow b$ ▷ Set a to b
5:          $b \leftarrow r$ ▷ Set b to r
6:          $r \leftarrow a \bmod b$ ▷ Update r
7:      **return** b ▷ The gcd is b

Exercise 1.13. Manually implement the Euclidean algorithm to find $\gcd(27, 63)$.

Solution.

(1) $a = 27$, $b = 63$.

(2) $r = a \bmod b = 27$

(3) $r \neq 0$

(4) $a = b = 63$

---

[4]A generalization of this procedure to polynomials is **polynomial long division**, which is described later in the book.

(5) $b = r = 27$

(6) ...

Remark 1.14. There is an alternative **recursive** version of the Euclidean algorithm. A function is called recursive if it calls itself during the execution. Below is pseudocode for the recursive version of Euclid's algorithm.

Algorithm Recursive Euclidean algorithm

1: **function** EUCLIDREC(a, b)
2:     $r \leftarrow a \bmod b$
3:     **if** $r = 0$ **then**
4:         **return** b
5:     **else**
6:         **return** EuclidRec(b, r)

We conclude this section with an example of an algorithm that involves nested *for* loops and conditional statements.

Example 1.15 (Finding substrings with specified first and last letters). For certainty, assume that all letter characters in the string are upper case. Suppose that a given string is split into a list S of strings of length one.[5] Finding all substrings of S starting with the letter "$L1$" and ending with the letter "$L2$" can then be implemented via the following algorithm:

Algorithm Find substrings

1: **function** SUBSTRINGS$(S, L1, L2)$
2:     $n \leftarrow length(S)$
3:     $M \leftarrow [\,]$ ▷ M is the empty set
4:     **for** $i \leftarrow 0, n - 2$ **do**   ▷ Loop variable i varies from 0 to $n - 2$
5:         **if** $S[i] \neq L1$ **then**
6:             $i \leftarrow i + 1$ ▷ If condition holds, update i
7:         **else**   ▷ If not, start the inner loop
8:             **for** $j \leftarrow i + 1, n - 1$ **do**   ▷ Loop variable j varies from $i + 1$ to $n - 1$
9:                 **if** $S[j] \neq L2$ **then**
10:                    $j \leftarrow j + 1$ ▷ If condition holds, update j
11:                **else**   ▷ If not,
12:                    $M.append\ S[i...j]$ ▷ Append $S[i...j]$ to M
13:     **return** M

💡 *Check Your Understanding.* In the above algorithm:

(a) Why does the largest value of i equal $n - 2$?

(b) Why does the largest value of j equal $n - 1$?

---

[5] Any general purpose CAS typically has a built-in function for this task.

To better understand how the algorithm is executed by a computer, let us manually implement the pseudocode to find all substrings starting with the letter $L1 = "A"$ and ending with $L2 = "B"$ in the string $"ABLEWAOBAK"$. We first write a list S of 10 substrings of length one:

$$S = ["A", "B", "L", "E", "W", "A", "O", "B", "A", "K"].$$

The execution of our algorithm is shown in the table below. The list of substrings M is originally empty, $M = [\ ]$.

i, S[i]	j, S[j]	M
$i = 0, S[i] = "A"$	$j = 1, S[j] = "B"$	$["AB"]$
	$j = 3, 4, 5, 6, S[j] \neq "B"$	
	$j = 7, S[j] = "B"$	$["AB", "ABLEWAOB"]$
	$j = 8, 9\ S[j] \neq "B"$	
$i = 1, 2, 3, 4, S[i] \neq "A"$		
$i = 5, S[i] = "A"$	$j = 6, S[j] \neq "B"$	
	$j = 7, S[j] = "B"$	$["AB", "ABLEWAOB", "AOB"]$
	$j = 8, 9, S[j] \neq "B"$	
$i = 6, 7, S[i] \neq "A"$		
$i = 8, S[i] = "A"$	$j = 9, S[j] \neq B$	$M = ["AB", "ABLEWAOB", "AOB"]$

Answer: $M = ["AB", "ABLEWAOB", "AOB"]$

One of the main concerns of the theory of algorithms is measuring the time needed to execute algorithms when the size n of the input grows. A mathematical tool called **big $\mathcal{O}$ notation** is crucial in classifying the time complexity of algorithms. Big $\mathcal{O}$ notation is written in the form $\mathcal{O}(f)$, where $\mathcal{O}$ stands for "order of magnitude" and f is a positive function that we use for comparing the behavior of a function or task of interest. In mathematics, the notation qualitatively describes the limiting behavior of a function g when the argument tends towards a particular value a or infinity. The formal definition of big $\mathcal{O}$ notation is somewhat delicate. Here we just state a sufficient condition for a function g to be in the class $\mathcal{O}(f)$ and illustrate this with a couple of examples. The condition reads

$$\text{If } \lim_{x \to a} |g(x)|/f(x) = c \text{ with } c > 0, \text{ then } g \in \mathcal{O}(f(x)) \text{ when } x \to a.$$

This statement is also true if $x \to \infty$.

Example 1.16.

(a) Show that $g(x) = 3x^2 + 5x - 1$ belongs to the class $\mathcal{O}(x^2)$ when $x \to \infty$.

(b) Show that $h(x) = \sin(2x)/x$ is of the class $\mathcal{O}(1)$ when $x \to 0$.

Solution.

(a) We have

$$\lim_{x \to \infty} \frac{3x^2 + 5x - 1}{x^2} = 3.$$

(b) Using L'Hopital's Rule and continuity of the cosine function, we obtain

$$\lim_{x \to 0} \frac{\sin(2x)}{x} = \lim_{x \to 0} \frac{2\cos(2x)}{1} = 2.$$

Exercise 1.17. Show that the function $f(x) = 1 - \cos(5x)$ belongs to the class $\mathcal{O}(x^2)$ when $x \to 0$.

In complexity theory, the function f that defines a complexity class of an algorithm depends on the size of the input $n \in \mathbb{N}$. Examples of complexity classes (in ascending order of complexity) are $\mathcal{O}(\log(n)), \mathcal{O}(n), \mathcal{O}(n\log(n)), \mathcal{O}(n^2)$, and $\mathcal{O}(2^n)$. For instance, the main operation in the algorithm of Example 1.6 is addition, and there are exactly n additions for an input of length n. Thus the algorithm belongs to the class $\mathcal{O}(n)$. Alternatively, we say that the complexity of the algorithm is linear. Similarly, the main operation in the algorithm of Example 1.8 is comparison, and there are $n - 1$ comparisons. This is also a case of linear time complexity.

1.4 Glossary

- **Computer algebra system** (CAS) – A mathematical software system that is capable of performing both symbolic and numeric calculations. It has the ability to manipulate mathematical expressions in a way similar to manual computations.

- **Algorithm** – A step-by-step procedure by which an operation can be carried out without any exercise of intelligence, and so, for example, by a machine.

- **Data type** – A classification of data which tells the compiler or interpreter what values and operations are acceptable for this data type. Examples are numeric types (real, integer, complex, Boolean) or composite types (set, list, array, dictionary). Composite data types are called **data structures**.

- **Identifier** – A name that is assigned to an object by the user.

- **Pseudocode** – A way to describe an algorithm using the structural conventions of formal programming language and mathematics, but with the intent for human reading rather than machine reading. It breaks an algorithm into one-sentence steps. No standard for pseudocode syntax exists and exactly how to structure pseudocode is a personal choice. However, the overriding goal should be clarity and precision.

- **Control flow statement** (in computer programming) – A statement that results in a choice being made as to which of two or more paths should be followed. Examples of control flow statements are loops and conditionals.

- **Modulo function** (or modulo operation) – In computing, the modulo function takes a pair m, n of positive integers and returns the remainder of dividing the first integer by the second. It is typically denoted by m mod n.

- **Recursive function** (in computer science) – A function that calls itself during the process of execution.

- **Big $\mathcal{O}$ notation** – A mathematical notation that qualitatively describes the limiting behavior of a function when the argument tends towards a specific value or infinity. It is written in the form $\mathcal{O}(f(x))$, where the letter $\mathcal{O}$ stands for "order of magnitude" and f represents a function used for comparing the behavior of a function or task of interest. In complexity theory of algorithms, the function f depends on the discrete argument $n \in \mathbb{N}$, the size of the input.

- **Time complexity** – A characteristic of an algorithm based on the number of elementary operations performed by the algorithm for a given input size. Time complexity is commonly measured using **big $\mathcal{O}$ notation**.

2

Topics in Classical Geometry

Classical geometry deals with geometric bodies and figures. It studies their mutual positions in three-dimensional space or on the plane, along with relevant magnitudes, such as distances, angles, lengths, and volumes. Coordinate geometry, also known as analytic geometry, uses a rectangular coordinate system in which each point is uniquely identified by a pair of numerical coordinates that represent the signed distance from the point to a set of perpendicular lines called coordinate axes. Analytic geometry was independently invented by René Descartes and Pierre de Fermat, but Descartes received all credit as its sole creator. In turn, the rectangular coordinate system is also called Cartesian, which comes from the Latin version of his last name Cartesius. **The Cartesian coordinate system changed mathematics in a profound way and led to the unification of geometry, algebra, and calculus.**

In this chapter we consider only two-dimensional geometric objects and a few problems of analytic geometry on the plane. Points on the plane admit another description – they can be represented as complex numbers. This representation leads to more clear and elegant proofs of some geometric results, as demonstrated by the examples, exercises, and lab on finding three special points of a triangle called the centroid, orthocenter, and circumcenter.

2.1 Review: Matrices, vectors, and lines

Matrix theory provides a great toolbox for mathematics and its applications. In this section we will introduce just a handful of terms and basic facts regarding small-size matrices in a somewhat informal language.

2.1.1 Matrices. A **matrix** of size $m \times n$ is a table of elements with m rows and n columns. Here are symbolic representations of 2×2 and 2×3 matrices:

$$A = \begin{bmatrix} a & b \\ c & d \end{bmatrix}, \qquad M = \begin{bmatrix} m_1 & m_2 & m_3 \\ m_4 & m_5 & m_6 \end{bmatrix}. \tag{2.1}$$

In this text we will be using only 2×2, 2×1, and 1×2 matrices. Below we provide quick explanations and examples of algebraic operations on such matrices.

Matrix addition. Matrices of the same size can be added element-wise. Here is an example.

Example 2.1. Let
$$B = \begin{bmatrix} 3 & 2 \\ -5 & 4 \end{bmatrix}$$
and A be the matrix defined above. Then
$$A + B = \begin{bmatrix} a+3 & b+2 \\ c-5 & d+4 \end{bmatrix}.$$

Exercise 2.2. Find the sum of the matrices B and
$$C = \begin{bmatrix} -1 & 0 \\ 7 & -5 \end{bmatrix}.$$

Scalar multiplication. Matrices can be multiplied by a scalar element-wise. Here is an example.

Example 2.3. Let A be the matrix in Equation (2.1) and k a real number. Then
$$kA = \begin{bmatrix} ka & kb \\ kc & kd \end{bmatrix}.$$

Exercise 2.4. Multiply matrix B in Example 2.1 by 3.

Remark 2.5. The difference $A - B$ of matrices of the same size can be calculated as the sum of the matrix A and the product $(-1)B$.

Matrix multiplication. Two matrices, A of size $m \times k$ and B of size $k \times n$, can be multiplied. Their product AB is a matrix of size $m \times n$. The general definition of the product of two matrices is somewhat cumbersome, so we define this operation here for only the three cases that we need in this text.

Definition 2.6.

(1) The product of a 1×2 matrix $A = [a_1 \ a_2]$ and 2×1 matrix $B = \begin{bmatrix} b_1 \\ b_2 \end{bmatrix}$ is the scalar
$$AB = a_1 b_1 + a_2 b_2. \tag{2.2}$$

(2) The product of a 2×2 matrix
$$A = \begin{bmatrix} a_1 & a_2 \\ a_3 & a_4 \end{bmatrix}$$
by a 2×1 matrix $B = \begin{bmatrix} b_1 \\ b_2 \end{bmatrix}$ is a 2×1 matrix defined by the following formula:
$$AB = \begin{bmatrix} a_1 b_1 + a_2 b_2 \\ a_3 b_1 + a_4 b_2 \end{bmatrix}. \tag{2.3}$$

(3) The product of two 2×2 matrices
$$A = \begin{bmatrix} a_1 & a_2 \\ a_3 & a_4 \end{bmatrix} \text{ and } B = \begin{bmatrix} b_1 & b_2 \\ b_3 & b_4 \end{bmatrix}$$
is a 2×2 matrix defined by the following formula:
$$AB = \begin{bmatrix} a_1 b_1 + a_2 b_3 & a_1 b_2 + a_2 b_4 \\ a_3 b_1 + a_4 b_3 & a_3 b_2 + a_4 b_4 \end{bmatrix}. \tag{2.4}$$

Example 2.7.

(a) The product of $A = \begin{bmatrix} 1 & -2 \end{bmatrix}$ and $B = \begin{bmatrix} 4 \\ 3 \end{bmatrix}$

$$AB = \begin{bmatrix} 1 & -2 \end{bmatrix} \begin{bmatrix} 4 \\ 3 \end{bmatrix} = -2.$$

(b) The product of

$$C = \begin{bmatrix} 1 & 0 \\ 2 & 3 \end{bmatrix} \text{ and } F = \begin{bmatrix} -1 \\ 4 \end{bmatrix}$$

is

$$CD = \begin{bmatrix} 1 & 0 \\ 2 & 3 \end{bmatrix} \begin{bmatrix} -1 \\ 4 \end{bmatrix} = \begin{bmatrix} -1 \\ 10 \end{bmatrix}.$$

(c) The product of

$$E = \begin{bmatrix} 7 & 5 \\ 2 & 3 \end{bmatrix} \text{ and } F = \begin{bmatrix} 2 & 1 \\ 5 & 4 \end{bmatrix}$$

is

$$EF = \begin{bmatrix} 7 & 5 \\ 2 & 3 \end{bmatrix} \begin{bmatrix} 2 & 1 \\ 5 & 4 \end{bmatrix} = \begin{bmatrix} 7 \cdot 2 + 5 \cdot 5 & 7 \cdot 1 + 5 \cdot 4 \\ 2 \cdot 2 + 3 \cdot 5 & 2 \cdot 1 + 3 \cdot 4 \end{bmatrix} = \begin{bmatrix} 39 & 27 \\ 19 & 14 \end{bmatrix}.$$

Exercise 2.8. Calculate the product FE for the matrices E and F defined in Example 2.7(c). Is the result the same as in the example?

Matrix transposition. Another important matrix operation is called **transposition**. The transpose operation swaps the rows and columns of a given matrix. For example, applied to the matrices A, B, and E in Example 2.7, the transposition operation returns the matrices

$$A^T = \begin{bmatrix} 1 \\ -2 \end{bmatrix}, \quad B^T = \begin{bmatrix} 4 & 3 \end{bmatrix}, \text{ and } E^T = \begin{bmatrix} 7 & 2 \\ 5 & 3 \end{bmatrix}.$$

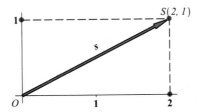

Figure 2.1. Vector **s** with its tail at the origin and head at the point $S(2, 1)$.

2.1.2 Plane vectors. The notion of a vector is a unifying concept between algebra and geometry. A geometric vector **s** is completely characterized by its magnitude (length) and direction. Vectors in geometry are called *free vectors* in the sense that any two geometric vectors of the same magnitude and direction are considered equivalent. In applications, geometric vectors are used as models of real-life quantities, like force and velocity.

When a rectangular coordinate system is chosen on the plane, a geometric vector can be defined by two numbers called the **coordinates of the vector**. Fig.2.1 shows an example of a geometric vector. The origin O is called the "tail" of the vector and the point S the "head". When the tail of a vector is at the origin, as is the case for the vector in Fig.2.1, the vector is called a **radius vector** (or **position vector**), and its coordinates are just the coordinates of the head. We will treat vectors in this book as 2×1 matrices or 1×2 matrices of their coordinates and call them **column vectors** or **row vectors**, respectively. So, the vector in Fig.2.1 can be written as the column vector $\mathbf{s} = \left[\begin{smallmatrix} 2 \\ 1 \end{smallmatrix} \right]$ or the row vector $\mathbf{s} = [2\ 1]$, depending on the context.

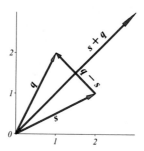

Figure 2.2. The sum and difference of vectors $\mathbf{q}$ and $\mathbf{s}$.

Since vectors can be treated as matrices, all matrix operations defined in this subsection for matrices, such as addition, scalar multiplication, and transposition, automatically apply to vectors. These algebraic operations can be visualized, as shown in Fig.2.2 for the sum and difference of two vectors.

☼ Check Your Understanding.

(a) Visually identify the coordinates of the vectors $\mathbf{q}$ and $\mathbf{s}$ in Fig.2.2, assuming that the coordinates are integers. Then find the coordinates of the vectors $\mathbf{q} + \mathbf{s}$ and $\mathbf{q} - \mathbf{s}$.

(b) What is the magnitude and direction of the vector $-2\mathbf{s}$ in terms of the length and direction of a vector $\mathbf{s}$?

A vector with tail at some point $M(x_1, y_1)$ and head at $M_2(x_2, y_2)$, denoted $\overrightarrow{M_1 M_2}$, has the coordinates $[x_2 - x_1, y_2 - y_1.]$ The **magnitude** $|\overrightarrow{M_1 M_2}|$ of this vector gives the distance between its endpoints and is calculated as

$$|\overrightarrow{M_1 M_2}| = \sqrt{(x_2 - x_1)^2 + (y_2 - y_1)^2}.$$

A very important vector operation called the **dot product** can be used to find the angle between two vectors.

Definition 2.9. The dot product of the vectors $\mathbf{q}$ and $\mathbf{s}$, denoted as $\mathbf{q} \cdot \mathbf{s}$, is the product of three numbers: the magnitudes of the vectors and the cosine of the angle between them. That is,

$$\mathbf{q} \cdot \mathbf{s} = |\mathbf{q}||\mathbf{s}| \cos(t),$$

where t is the angle between the two vectors.

Notice that if the dot product of two nonzero vectors[1] is zero, then the vectors are perpendicular. This definition implies that the angle t between the vectors $\mathbf{q}$ and $\mathbf{s}$ can be found by the formula

$$\arccos(t) = \frac{\mathbf{q} \cdot \mathbf{s}}{|\mathbf{q}||\mathbf{s}|}.$$

It is known that if $\mathbf{q} = [q_1 \ q_2]$ and $\mathbf{s} = [s_1 \ s_2]$, then the dot product can be calculated as

$$\mathbf{q} \cdot \mathbf{s} = q_1 s_1 + q_2 s_2.$$

Dotting a vector with a **unit vector**, that is, a vector of length one, yields the **scalar projection** of the given vector onto the direction of the unit vector. If the vector $\overrightarrow{OA}$ in Fig. 2.3 has magnitude one, the scalar projection of the vector $\overrightarrow{OB}$ onto the direction of $\overrightarrow{OA}$ is the length of the vector $\overrightarrow{OP}$. Note that the scalar projection can be negative. The vector $\overrightarrow{OP}$ is called the **vector projection** of the vector $\overrightarrow{OB}$ onto the direction of the vector $\overrightarrow{OA}$.

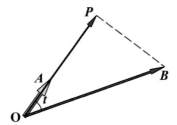

Figure 2.3. Vector $\overrightarrow{OP}$ is the **vector projection** of $\overrightarrow{OB}$ onto the direction of the vector $\overrightarrow{OA}$. If $|\overrightarrow{OA}| = 1$, the **scalar projection** of $\overrightarrow{OB}$ on $\overrightarrow{OA}$ is the dot product $\overrightarrow{OA} \cdot \overrightarrow{OB}$. In this case, it is just the length of the vector projection $\overrightarrow{OP}$.

Exercise 2.10. Make a CAS function *angle_and_projection*($\mathbf{v}_1, \mathbf{v}_2$) that takes two vectors and returns

- The angle between the vectors $\mathbf{v}_1$ and $\mathbf{v}_2$, in radians.
- A unit vector in the direction of the vector $\mathbf{v}_2$.
- the scalar projection of $\mathbf{v}_1$ onto the direction of $\mathbf{v}_2$.

Test your function on an example of your choice.

2.1.3 Lines. A line on the plane can be defined by:

- The slope m and y-intercept b as $y = mx + b$. (This format applies only to non-vertical lines. **Why?**)

- A point (x_0, y_0) on the line and a normal (perpendicular) vector $\mathbf{n} = [a \ b]$ as the so called **general equation of the line** $a(x - x_0) + b(y - y_0) = 0$, or $ax + by + c = 0$ with $c = -ax_0 - by_0$. We can interpret the left-hand side of the general equation as the dot product of a normal vector and a variable vector $[x - x_0 \ y - y_0]$ collinear to the line.

---

[1]The zero vector is the vector $[0 \ 0]$.

Figure 2.4. The distance from the point A to the solid line is the absolute value of the dot product of the vector $\overrightarrow{AB}$ and the unit normal **n**.

- A point (x_0, y_0) on the line and a direction vector $\mathbf{s} = [s_1\ s_2]$ as $(x - x_0)/s_1 = (y - y_0)/s_2$.

Observe that:

(1) The equation $(x - x_0)/s_1 = (y - y_0)/s_2$ can be rewritten as $s_2(x - x_0) - s_1(y - y_0) = 0$, which means that if $\mathbf{s} = [s_1\ s_2]$ is a direction vector of a line, then the vector $\mathbf{n} = [s_2\ -s_1]$ is a normal vector to the line.

(2) Since the coordinates of the variable vector are proportional to the coordinates of the direction vector of the line, we can introduce a new variable t, called the coefficient of proportionality, and obtain the **parametric equations of the line** $x = x_0 + ts_1, y = y_0 + ts_2$.

(3) The angle between two lines can be found as the angle between normal vectors to the lines.

In the next example we will show how the objects and operations from this brief review work by deriving the formula for the distance between a point and a line.

Example 2.11. Consider a line l defined by the general equation $ax + by + c = 0$. Let $A(x_0, y_0)$ be some point. Derive the formula for the distance from A to the given line.

Solution. Let $B(x, y)$ be an arbitrary point on l. The distance d from A to the line can be found as the absolute value of the scalar projection of the vector $\overrightarrow{AB} = [x - x_0\ y - y_0]$ onto the unit normal vector $\mathbf{n} = [a\ b]/\sqrt{a^2 + b^2}$ (see Fig. 2.4):

$$d = |\overrightarrow{AB}\cdot\mathbf{n}| = \frac{|(x - x_0)\,a + (y - y_0)\,b|}{\sqrt{a^2 + b^2}}.$$

Expanding the numerator and using the fact that point B lies on the line, we arrive at the formula

$$d = \frac{|ax_0 + by_0 + c|}{\sqrt{a^2 + b^2}}. \tag{2.5}$$

💡 **Check Your Understanding.** Show the details of the last step in the derivation of this formula.

Exercise 2.12.

(a) Make a CAS function *pt_line_distance*($\mathbf{a}, \mathbf{b}, \mathbf{c}, \mathbf{pt}$) that takes coefficients a, b, c of the general equation of a line and coordinates $pt = (x_0, y_0)$ of a point and returns the distance from the point to the line.

(b) Test your function on an example of your choice. Use CAS to plot (i) the point, (ii) the line, and (iii) a vector perpendicular to the line with its tail located at the given point in one figure. Your figure will be similar to Fig. 2.4.

2.2 Rigid transformations of the plane

A **rigid transformation** is a one-to-one map of the plane onto itself that preserves the distance between every pair of points. Parallel translations, rotations about the origin, reflections about the coordinate axes, and all possible compositions of these rigid transformations constitute the set of all rigid transformations of the plane.[2] In the following we use matrices and radius vectors of points to simplify notation.

- A **parallel translation** $P : [x\ y]^T \mapsto [x + a\ y + b]^T$ maps the vector $\mathbf{v}_{old} = \begin{bmatrix} x \\ y \end{bmatrix}$ to $\mathbf{v}_{new} = \mathbf{v}_{old} + \mathbf{s}$, where the vector $\mathbf{s} = \begin{bmatrix} a \\ b \end{bmatrix}$ specifies the translation. In vector notation we write $\mathbf{v}_{new} = P(\mathbf{v}_{old})$.

- A **rotation by an angle** t **about the origin**

$$R : [x\ y]^T \mapsto [x \cos(t) + y \sin(t)\ -x \sin(t) + y \cos(t)]^T$$

can be defined in matrix notation as $\mathbf{v}_{new} = A\mathbf{v}_{old}$. Here $\mathbf{v}_{old} = \begin{bmatrix} x \\ y \end{bmatrix}$ and the **rotation matrix** A is defined as

$$A = \begin{bmatrix} \cos(t) & \sin(t) \\ -\sin(t) & \cos(t) \end{bmatrix}. \tag{2.6}$$

Remark 2.13. Positive and negative t-values specify counterclockwise and clockwise rotations, respectively. Therefore $\mathbf{v}_{old} = A^{-1}\mathbf{v}_{new}$, where

$$A^{-1} = \begin{bmatrix} \cos(t) & -\sin(t) \\ \sin(t) & \cos(t) \end{bmatrix}$$

is the matrix that defines a rotation by the angle $-t$. It is the **inverse** of the matrix A, that is,

$$AA^{-1} = A^{-1}A = I \equiv \begin{bmatrix} 1 & 0 \\ 0 & 1 \end{bmatrix}. \tag{2.7}$$

Thus, the inverse of the rotation matrix A is just the transposition of A, that is, $A^{-1} = A^T$.

☼ **Check Your Understanding.** Verify formula (2.7).

- A **reflection about the** x**-axis** $F : [x\ y]^T \mapsto [x\ -y]^T$ can be defined in matrix notation as $\mathbf{v}_{new} = F\mathbf{v}_{old}$ with $F = \begin{bmatrix} 1 & 0 \\ 0 & -1 \end{bmatrix}$. A reflection transformation over the y-axis can be defined similarly. (**Do this!**)

---

[2] A composition of two transformations is a consecutive implementation of them.

In the next exercise an important property of compositions of rigid transformations is analyzed. Compositions of two rotations, two reflections, and rotations and reflections are defined by the product of matrices corresponding to these transformations. A composition of a translation with a rotation or reflection is just a consecutive implementation of these transformations, which can be written using the notation introduced above. For example, the image of rotating and then translating a vector $\mathbf{v}$ is the vector $P(A\mathbf{v})$.

Exercise 2.14. Show the following.

(a) The result of two consecutive rotations of the plane defined by the rotation matrices A_1 and A_2 does not depend on the order of these transformations. That is, $A_1A_2\mathbf{s} = A_2A_1\mathbf{s}$. We say in this case that the two transformations **commute**.

(b) Generally, rotation and reflection transformations do not commute. That is, $AF\mathbf{s} \neq FA\mathbf{s}$.

(c) Generally, rotations and translations do not commute. That is, $AP(\mathbf{s}) \neq P(A\mathbf{s})$.

(d) Generally, reflections and translations do not commute. That is, $FP(\mathbf{s}) \neq P(F\mathbf{s})$.

Suggestions: Do part (a) in symbolic form manually or with CAS assistance. (Any general purpose CAS provides vector and matrix operations. Just define the symbolic matrices for rotation and reflection and vector for translation and compute appropriate compositions.) For parts (b) through (d) it suffices to give a concrete examples of compositions that do not commute.

Remark 2.15. Rigid transformations of three-dimensional space are defined similarly, with rotations and reflections represented by appropriate 3×3 matrices and translations defined by three-dimensional vectors. As opposed to rotations of the plane, rotations in three-dimensional space generally do not commute.

A set of translations and rotations in $\mathbb{R}^3$ and their compositions, equipped with the composition operation (consecutive implementation of these transformations), forms an algebraic structure known as a **special Euclidean group**. In general, a group is a set of mathematical objects equipped with a **binary operation** satisfying certain axioms. For any pair of elements in the group, this binary operation returns a certain element also belonging to the group. If the result of the operation does not depend on the order of the two inputs, the group is called **abelian** or **commutative**. A group is one of the basic mathematical structures studied in abstract algebra courses.

Exercise 2.16. Make a CAS function ***three_transformations*(pt, t, v, p)** that takes coordinates *pt* of a point, angle t (in radians), vector $\mathbf{v}$, and parameter p. The function should return a figure with the given point and its image under the following transformations:

(a) Translation defined by the vector $\mathbf{v}$ if $p = 0$. Include the plot of vector $\mathbf{v}$ with tail at the given point in your figure.

(b) Rotation defined by the angle t if $p = 1$. Include the trajectory of the rotating point in your figure.

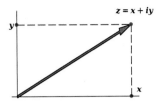

Figure 2.5. Complex number $x + iy$ representing the radius vector $[x\ y]$.

(c) Reflection of the point about the y-axis if $p = 2$. Include the horizontal segment connecting the given point and its image in your figure.

Test the function for each of the three kinds of motions on examples of your choice. **Suggestion**: Use an appropriate CAS conditional statement *if-elif-else* to control the flow of the function implementation for each of the three possible values of the parameter p.

There are branches of geometry different from Euclidean geometry. Projective geometry, Riemannian geometry, and Lobachevskian (hyperbolic) geometry are examples of non-Euclidean geometries. Each studies properties of geometric objects that remain unchanged (**invariant**) under certain classes of transformations.

2.3 Complex numbers in classical geometry

This subsection is just a glimpse of the complex language in geometry. Representing a point or its radius vector by a complex number allows us to easily prove some classical geometric facts.

2.3.1 About complex numbers. Complex numbers can be defined as ordered pairs (x, y) of real numbers equipped with addition and multiplication operations that satisfy certain rules. Perhaps a more familiar form of a complex number is $z = x + iy$, where $i \equiv \sqrt{-1}$ is called the **imaginary unit**. The real numbers x and y are called the real and imaginary parts of the complex number z, respectively, and we write $x = \mathrm{Re}(z)$, $y = \mathrm{Im}(z)$. Any real number is a complex number with imaginary part zero. A complex number can be visualized as a point on the plane or the radius vector of this point (see Fig. 2.5). The **complex conjugate** of $z = x + iy$ is defined to be $z^* = x - iy$. Recall that the sum and the product of two complex numbers $z = x + iy$ and $w = u + iv$ are $(x + u) + i(y + v)$ and $(x \cdot u - y \cdot v) + i(x \cdot v + y \cdot u)$, respectively. Notice that $z \cdot z^* = r^2$, where $r = |z|$ is the length of the radius vector representing z. The reciprocal $1/z$ of a complex number $x + iy$ is the number $x/(x^2 + y^2) - y/(x^2 + y^2)$. (**Show this!**)

If we identify any row vector $[x\ y]$ or column vector $\left[\begin{smallmatrix} x \\ y \end{smallmatrix}\right]$ with the complex number $x + iy$, then scalar multiplication, addition, and subtraction of vectors can be done on corresponding complex numbers. For example, consider a row vector $\overrightarrow{AB}$ with the point A represented by $z = x + iy$ and the point B represented by $w = u + iv$. Then $\overrightarrow{AB} = [u - x, v - y]$, which corresponds to the difference $w - z$. Alternatively, we can interpret this as the difference of two radius vectors: $\overrightarrow{AB} = \overrightarrow{OB} - \overrightarrow{OA}$.

Exercise 2.17.

(a) Let the vectors $\mathbf{q} = [x\ y]$ and $\mathbf{s} = [u\ v]$ represent the points $z = x + iy$ and $w = u + iv$. Show that the dot product of the vectors $\mathbf{q}$ and $\mathbf{s}$ is equal to the real part of the product of z and the complex conjugate of w. That is, $\mathbf{q} \cdot \mathbf{s} = \operatorname{Re}(z \cdot w^*)$.

(b) Show that if a vector $\mathbf{v} = [v_1\ v_2]$ is represented by a complex number $z = v_1 + iv_2$, then the complex number iz represents a vector orthogonal to $\mathbf{v}$.

The position of a point $P(x, y)$ with $x^2 + y^2 > 0$ is completely defined by the distance $r = \sqrt{x^2 + y^2}$ of the point from the origin and the angle φ, $\varphi \in (-\pi, \pi]$, subtended between the positive real axis and the radius vector $\overrightarrow{OP}$. The tuple (r, φ) is called the **polar coordinates** of P. For the corresponding complex number $z = x + iy$, r is the **modulus** and φ is the **argument** of z. (In applications the argument is often called the **phase**.) The polar coordinates r and φ can be converted to the rectangular coordinates x and y using the formulas $x = r \cos \varphi$ and $y = r \sin \varphi$. It follows that a complex number $z = x + iy$ can be written in **polar form** as $z = r(\cos(\varphi) + i \sin(\varphi))$.

Exercise 2.18. Make a function *find_argument*($\mathbf{a}, \mathbf{b}$) that takes the real and imaginary parts of a complex number and returns the argument of the number. Test the function on a complex number of your choice.

2.3.2 Euler's formula. The Nobel prize winning physicist Richard Feynman called Euler's formula "our jewel" and "the most remarkable formula in mathematics".[3] Here is this mathematical jewel: For any real number t,

$$e^{it} = \cos t + i \sin t. \tag{2.8}$$

It follows that $|e^{it}| = 1$. We will take formula (2.8) as the definition of e^{it}. This formula implies that

$$\cos t = \frac{e^{it} + e^{-it}}{2} \quad \text{and} \quad \sin t = \frac{e^{it} - e^{-it}}{2i}. \tag{2.9}$$

An exponential function of a complex argument $z = x + iy$ can be defined as

$$e^z = e^x e^{iy} = e^x(\cos y + i \sin y).$$

To simplify notation we will write e^z as $\exp(z)$. Clearly, the modulus $|\exp(z)|$ of the complex exponential function equals $\exp(x)$. (**Show this!**) The complex exponential function preserves certain properties of its real cousin, in particular, $\exp(z) \cdot \exp(w) = \exp(z + w)$. Using Euler's formula, the polar form of a complex number can be written as $z = r \exp(i\varphi)$.

🔆 **Check Your Understanding.** Let $z_j = r_j \exp(i\varphi_j)$, $j = 1, 2$. Show that

$$z_1 z_2 = r_1 r_2 (\cos(\varphi_1 + \varphi_2) + i \sin(\varphi_1 + \varphi_2)).$$

Remark 2.19. If a point X on the plane is represented by a complex number $r \exp(i\varphi)$, the product of this number and $\exp(it)$, a number with modulus one, represents the image of the rotation of X about the origin through the angle t. Thus, each point on the

---

[3]The Feynman Lectures on Physics, Volume I, Ch 22.

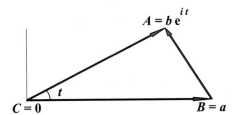

Figure 2.6. A triangle with vertices in special positions represented by complex numbers.

unit circle $|z| = 1$ specifies a rotation transformation and therefore can be represented by the matrix $A_z = \begin{bmatrix} \cos(t) & \sin(t) \\ -\sin(t) & \cos(t) \end{bmatrix}$. This justifies the representation of any complex number $z = a + bi$ by the matrix $A_z = \begin{bmatrix} a & b \\ -b & a \end{bmatrix}$ (**explain why**).

Exercise 2.20. Show that if the matrices A_z, A_w represent complex numbers z, w, then the matrices $A_z + A_w$, $A_z A_w$ represent the sum $z + w$ and the product $z \cdot w$, respectively.

Exercise 2.20 suggests that two different mathematical spaces, the set of complex numbers and the set of real 2×2 matrices of the form $\begin{bmatrix} a & b \\ -b & a \end{bmatrix}$, both equipped with the standard addition and multiplication operations, are in some (rigorous algebraic) sense "the same".

2.3.3 Examples of using complex numbers in geometry.

Example 2.21 (Derivation of the Law of Cosines). Consider the triangle ABC in Fig. 2.6 with side lengths a, b, c, where c is the length of vector $\overrightarrow{BA}$. The Law of Cosines claims

$$c^2 = a^2 + b^2 - 2\,ab \cos(t). \tag{2.10}$$

The vector $\overrightarrow{BA}$ is the difference of the vectors $\overrightarrow{CA} = b \exp(it)$ and $\overrightarrow{CB} = a$, that is, $\overrightarrow{BA} = b \exp(it) - a$. Since for any complex number z, $|z|^2 = zz^*$, we have

$$c^2 = \big(b \exp(it) - a\big) \cdot \big(b \exp(-it) - a\big).$$

After some elementary algebraic manipulations, we obtain

$$c^2 = a^2 + b^2 - ab\big(\exp(it) + \exp(-it)\big) = a^2 + b^2 - 2ab \cos(t). \quad \square$$

Exercise 2.22. Make a function *find_angle*(L) that takes a list $L = [a, b, c]$ of side lengths of a triangle and finds the angle between the sides of lengths a an b. Choose a triangle with known side lengths and use the function to find all three angles of the triangle.

Remark 2.23. To make the exercise more challenging, the reader can include in the function a verification that a triangle with the given side lengths exists. For example, the triangle with side lengths $13, 17, 3$ does not exist. (**Think why or just try to plot a triangle with these side lengths.**)

Example 2.24 (Complex form of parametric equation of the line passing through two given points). Let $A(x_1, y_1), B(x_2, y_2)$ be two points. We identify them with the complex numbers $z_j = x_j + iy_j, j = 1, 2$. Then $z_2 - z_1$ is a direction vector of the line passing through these points. Let z be a generic point on the line. Then a parametric equation of the line can be written in the complex form as

$$z = z_1 + t(z_2 - z_1).$$

Notice that the two scalar equations in the conventional parametric equations of a line are nicely combined into one equation when using complex numbers. This form simplifies the proof of the theorem in the next subsection about the **centroid** of a triangle.

2.4 Three centers of a triangle

2.4.1 The centroid.
Recall that a median of a triangle is a line segment connecting a vertex to the midpoint of the opposite side.

Theorem 2.25. *The three medians of any triangle intersect at one point called the centroid.*[4] *See Fig. 2.7.*

Below we provide a partial proof of the theorem.

Proof. Without loss of generality, assume that a triangle is positioned as in Fig 2.7. Our plan is to write the equations of the three medians of this triangle in complex parametric form and then show that the system of these three equations has a unique solution, the centroid.[5]

For simplicity, we denote the triangle's vertices by $A_j, j = 1, 2, 3$, and identify them with appropriate complex numbers so that $A_1 = 0$, $A_2 = 2a$, $A_3 = 2c_1 + 2ic_2$. Recall that the midpoint of a segment with endpoints z_1 and z_2 is $(z_1 + z_2)/2$. Thus, the midpoints of the triangle sides are

$$M_1 = a + c_1 + ic_2,$$
$$M_2 = c_1 + ic_2,$$
$$M_3 = a.$$

Next, we need direction vectors for the medians of the triangle $A_1 A_2 A_3$. These are the vectors

$$\mathbf{s}_1 = \overrightarrow{OM_1} = a + c_1 + ic_2,$$
$$\mathbf{s}_2 = \overrightarrow{OM_2} - \overrightarrow{OA_2} = c_1 + ic_2 - 2a,$$
$$\mathbf{s}_3 = \overrightarrow{OM_3} - \overrightarrow{OA_3} = a - 2(c_1 + ic_2).$$

Using the format $z = A_j + t_j \cdot \mathbf{s}_j$ for the equation of a median passing through the vertex A_j, we obtain the parametric equations of the medians in the complex form

$$z = t_1(a + c_1 + ic_2) \text{ for } A_1 M_1,$$
$$z = 2a + t_2(c_1 + ic_2 - 2a) \text{ for } A_2 M_2, \tag{2.11}$$
$$z = a + t_3(a - 2c_1 - 2ic_2) \text{ for } A_3 M_3,$$

---

[4]In physics the centroid is called the "center of mass".
[5]The last step will be done in Exercise 2.26.

where t_j, $j = 1, 2, 3$, are parameters. To complete the proof we have to find parameter values t_1^*, t_2^*, t_3^* that make the right-hand sides of these equations equal. This will be done in the next exercise.

Exercise 2.26. Denote the right-hand sides of the equations in (2.11) as $f_j(t_j)$ and find the solution t_1, t_2, t_3 of the system $f_1(t_1) = f_2(t_2)$, $f_2(t_2) = f_3(t_3)$.
Suggestion: Each of these two complex equations is equivalent to a system of two linear equations with respect to the parameters t_j. Write each of the systems in real form and solve it manually or with CAS assistance.

It turns out that the values of these parameters are equal to the same number $t^* = 2/3$. Thus, the centroid exists and its distance from any vertex is two-thirds of the length of the corresponding median. $\square$

As a bonus of our hard work, we have rediscovered the following fact: *For any triangle, the centroid divides each median in a ratio of 2 to 1.* This fact makes finding the centroid of any triangle very easy: choose any vertex and find the sum of the radius vector of the vertex and 2/3 times the vector representing the median from this vertex.

💡 **Check Your Understanding.** Why is this fact that we discovered for a triangle at a special position true for any triangle?

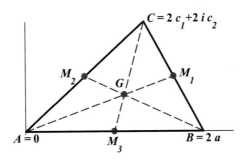

Figure 2.7. The centroid of a triangle is the point of intersection of its medians.

2.4.2 Lab 1: Three centers of a triangle at a special position. It is known that the three altitudes of any triangle intersect at one point called the **orthocenter**. Also, the three perpendicular bisectors of the sides of any triangle intersect at the center of the circumscribed circle called the **circumcenter**.
    The goal of this lab is to find the three centers of a triangle (the centroid, orthocenter, and circumcenter) when the triangle is located at a special position.

Problem formulation. Consider a triangle with vertices $A(0,0)$, $B(a,0)$, $C(c_1, c_2)$. Find the centroid, orthocenter, and circumcenter of the triangle. Choose specific values of a, c_1 c_2 and illustrate each center with a figure that includes a triangle, segments of the lines used for finding the center, and the center itself. Also, design an isosceles right triangle with hypotenuse *AB* and find the three centers of the triangle. (Note that proving the existence of the orthocenter and circumcenter is not required. To find these centers just find the intersections of any two altitudes and any two bisectors of the sides of the given triangle.)

Suggested plan and directions.

Part 1.

- Make a function ***centroid_special*(pt1, pt2, pt3)** that takes coordinates of the vertices of a triangle and returns the coordinates of the centroid.
- Make a function ***plot_centroid*(pt1, pt2, pt3)** that calls the function ***centroid_special*(pt1, pt2, pt3)** and makes the required figure. Test your function on an example of your choice. Your figure will be similar to Fig. 2.8.

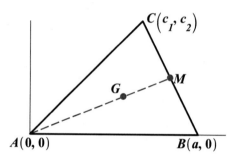

Figure 2.8. The centroid cuts the median into two segments whose lengths are in a 2:1 ratio.

Part 2.

- Make a CAS function ***ortho_special*(pt1, pt2, pt3)** that returns the coordinates of the orthocenter.
- Make a function ***plot_ortho*(pt1, pt2, pt3)** that calls the function ***ortho_special*(pt1, pt2, pt3)** and returns the required figure. Test your function on an example of your choice. Your figure will be similar to Fig. 2.9

Note that the equation of the vertical altitude is trivial. Use the general equation of a line to find the second altitude. Use an appropriate CAS command for finding the intersection point of the altitudes.

Part 3.

- Make a function ***cCenter_special*(pt1, pt2, pt3)** that returns coordinates of the circumcenter.
- Make a function ***plot_cCenter*(pt1, pt2, pt3)** that calls the function ***cCenter_special*(pt1, pt2, pt3)** and makes the required figure. Test your function on an example of your choice. Your figure will be similar to Fig. 2.10.

Part 4. Hint for constructing the isosceles right triangle: To find the coordinates of the vertex *C*, imagine that the triangle is inscribed in a circle and recall how inscribed angles are measured.

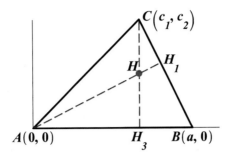

Figure 2.9. The orthocenter of a triangle is the point of intersection of its altitudes.

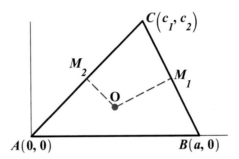

Figure 2.10. The circumcenter of a triangle is the point of intersection of its bisectors.

Part 5. Run each of the functions for finding the three centers of a triangle on the symbolic input $pt1(0,0)$, $pt2(a,0)$, $pt3(c1,c2)$. The outputs give explicit formulas for the three centers of a triangle with the given symbolic vertices. Note that the formula for the coordinates of the centroid is nicely related to the vertices' coordinates. Write a complete sentence about this relation.

The three formulas obtained in the last part of this lab will be used in Lab 2 of the next chapter in testing the functions that find the three centers of a triangle in a generic position.

2.5 Glossary

- **Matrix of size m×n** – A rectangular table of elements with m rows and n columns. A matrix with $m = n$ is called square.

- **Vector** – A geometric vector is completely characterized by its magnitude (length) and direction. A **unit vector** is a vector of length one. A plane vector (vector on the coordinate plane) is described by its coordinates that can be written as a row vector $[v_1 \ v_2]$ or as a column vector $\begin{bmatrix} v_1 \\ v_2 \end{bmatrix}$.

- **Dot product** of two vectors – The product of the magnitudes of the vectors and the cosine of the angle between them.

- **Rigid transformations** – Translations, rotations, reflections, and compositions of these transformations.

- **Polar coordinates of a complex number** $z = x + iy$ – The ordered pair (r, φ), where $r = \sqrt{x^2 + y^2}$ is the **modulus** of z and $\varphi \in (-\pi, \pi]$ the angle defined by the conditions $\cos(\varphi) = x/r$, $\sin(\varphi) = y/r$, is the **argument** of z.

- **Euler's formula**: $e^{it} = \cos t + i \sin t$.

- **Centroid of a triangle** – The intersection point of the three medians of a triangle.

- **Orthocenter of a triangle** – The intersection point of the three altitudes of a triangle.

- **Circumcenter of a triangle** – The intersection point of the three perpendicular bisectors of a triangle.

3

More Topics in Classical Geometry

In this chapter some basic tools described in the previous chapter will be put to work in two labs based on the Euler line and Simson line. Then, after a brief review of equations of conics and some transformations of these equations, we suggest applying appropriate techniques for certain related tasks.

3.1 Lab 2: The Euler line

In Chapter 3 we introduced three special points of a triangle called the centroid, orthocenter, and circumcenter, and found these points for a triangle located at a special position. It is known that these three special points of any non-equilateral triangle are **collinear**. The points lie on a line called the **Euler line**. The goal of this lab is to constructively prove that the Euler line exists, find the equation of this line, and plot it.

Problem formulation. Given three points, $pt1$, $pt2$, $pt3$, check if the points are collinear or if the triangle with vertices at these points is equilateral. If neither of these two conditions is true, find the centroid, the orthocenter, and the circumcenter of the triangle. Prove that the Euler line exists and find its equation. Then plot the triangle, the three centers, and the Euler line in one figure. Your figure will be similar to Fig. 3.1.

Suggested plan and directions.

Part 1. Make four CAS functions:

- *check_data*(**pt1, pt2, pt3**) that takes coordinates of three points and returns the number 1 if one of the two conditions (the triangle with vertices at these points is equilateral or the points are collinear) is true and 0 otherwise. Note that three points T_1, T_2, T_3 are collinear if and only if the sum of the vectors $\overrightarrow{T_1 T_2}$ and $\overrightarrow{T_2 T_3}$ equals $\overrightarrow{T_1 T_3}$.

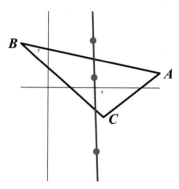

Figure 3.1. A triangle, its three special centers, and the Euler line.

Suggestion: You can check these two conditions separately using an *if-elif-else* conditional statement or construct a *compound* condition[1] and use an *if-else* statement instead.

- *centroid*(**pt1, pt2, pt3**) that returns the coordinates of the centroid
- *orthocenter*(**pt1, pt2, pt3**) that returns the coordinates of the orthocenter
- *circumcenter*($pt1, pt2, pt3$) that returns the coordinates of the circumcenter.

Test these functions using the symbolic input $pt1(0,0)$, $pt2(a,0)$, $pt3(c1,c2)$. The coordinates of the three centers are known from Lab 1 in Chapter 3.

Part 2. It suffices to prove the existence of the Euler line for a triangle with vertices $A(0,0)$, $B(a,0)$, $C(c1,c2)$.

Part 3. Make a function *euler_line*(**pt1, pt2, pt3**) that calls the function *check_data*(**pt1, pt2, pt3**) and halts if the output is 1. Otherwise, it calls the other three functions made in Part 1 and returns

- the coordinates of the three centers of the triangle
- the equation of the Euler line
- a plot of the triangle, the three centers, and the Euler line in one figure

Test your function on a triangle in a generic position. Your figure will be similar to Fig. 3.1.

Part 4. Let the point $pt1$ be located at the origin. Describe all pairs of points $pt2$, $pt3$ such that the triangles with vertices $pt1$, $pt2$, $pt3$ have horizontal Euler lines. Pick one of these triangles and use the function made in the previous part to make a corresponding plot.

💡 **Check Your Understanding.** The blue points in Fig. 3.1 are the center of mass, the orthocenter, and the circumcenter of the depicted triangle. Identify which is which. Use the words "highest point", "middle point", and "lowest point" in your answer.

---

[1]See Ch. 1 for a simple example.

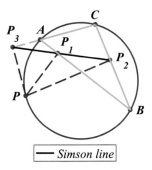

Figure 3.2. Simson line.

💡 **Check Your Understanding.** For a generic triangle the Euler line is unique. An equilateral triangle does not have a unique Euler line. Why?

📖 The Euler line also passes through two other special points of a triangle, the center of the so called **nine-point circle** and the **Exeter point**.

3.2 The Simson line

Consider a triangle ABC and an arbitrary point P on its circumcircle (see Fig. 3.2). It is known that the three points on the sides of the triangle or on their extensions closest to P are collinear. Clearly, these closest points are the intersections of the perpendiculars dropped from the point P to the sides of the triangle or their extensions. We will call these points "feet of perpendiculars". Since the feet of the perpendiculars are collinear, there is a line passing through them called the Simson line. This section contains an existence theorem and proof for the Simpson line and a lab. Some of the claims in the theorem are suggested as exercises. The lab includes deriving the equation of the Simson line and plotting it.

3.2.1 Theorem on existence of the Simson line.

Theorem 3.1. *The feet of the perpendiculars from an arbitrary point on the circumcircle of a triangle to the sides of the triangle or their extensions are collinear.*

Proof. To simplify derivations, we assume without loss of generality that the center of the circumscribed circle of a triangle is at the origin. Using an appropriate *uniform scaling*[2] if needed, we can also assume that the radius of the circle is one. This assumption means that the vertices of our triangle lie on the unit circle. Let $A(\exp(it_1))$, $B(\exp(it_2))$, $C(\exp(it_3))$ be the positions of the vertices. Let $P(\exp(it_0))$ be an arbitrary point on the circle that is different from the vertices.

First, we find the feet of the three perpendiculars from P to the sides of the triangle and then show that these points are collinear. To reduce technicalities, we find a symbolic solution of the following problem: *Given three points, $M_1(\exp(i\alpha))$, $M_2(\exp(i\beta))$ and $P(\exp(it_0))$, find the foot of the perpendicular dropped from P on the line M_1M_2.*

---

[2] A uniform scaling is a linear transformation that enlarges or shrinks object by a scale factor that is the same in all directions. It is known that a uniform scaling changes distances but does not change angles.

A direction vector of the line $M_1 M_2$ is $\mathbf{q} = [\cos(\beta) - \cos(\alpha) \ \sin(\beta) - \sin(\alpha)]$. We can use a simpler parallel vector $\mathbf{s} = \left[-\sin\left(\frac{\alpha+\beta}{2}\right) \ \cos\left(\frac{\alpha+\beta}{2}\right) \right]^T$.

💡 **Check Your Understanding.** Show that the vectors $\mathbf{q}$ and $\mathbf{s}$ are parallel. **Hint:** Use the trig identities

$$\cos(\alpha) - \cos(\beta) = -2\sin\left(\frac{\alpha+\beta}{2}\right)\sin\left(\frac{\alpha-\beta}{2}\right),$$

$$\sin(\alpha) - \sin(\beta) = 2\cos\left(\frac{\alpha+\beta}{2}\right)\sin\left(\frac{\alpha-\beta}{2}\right).$$

Now the general equations of the line $M_1 M_2$ and the perpendicular from P to this line can be written as

$$\cos\left(\frac{\alpha+\beta}{2}\right)(x - \cos(\alpha)) + \sin\left(\frac{\alpha+\beta}{2}\right)(y - \sin(\alpha)) = 0,$$
$$\sin\left(\frac{\alpha+\beta}{2}\right)(x - \cos(t_0)) - \cos\left(\frac{\alpha+\beta}{2}\right)(y - \sin(t_0)) = 0. \tag{3.1}$$

This is a system of linear equations. There is a nice explicit formula for solving such linear systems called Cramer's Rule.[3] For a linear system

$$a_1 x + b_1 y = c_1,$$
$$a_2 x + b_2 y = c_2, \tag{3.2}$$

Cramer's rule gives the unique solution $x = \Delta_x/\Delta$, $y = \Delta_y/\Delta$, where

$$\Delta = \det\begin{bmatrix} a_1 & b_1 \\ a_2 & b_2 \end{bmatrix}, \ \Delta_x = \det\begin{bmatrix} c_1 & b_1 \\ c_2 & b_2 \end{bmatrix}, \ \Delta_y = \det\begin{bmatrix} a_1 & c_1 \\ a_2 & c_2 \end{bmatrix}$$

provided $\Delta \neq 0$.

Rewriting the system (3.1) in the form (3.2) and solving it using Cramer's rule, we obtain (after some tedious, but elementary trigonometric manipulations)

$$x = \frac{1}{2}(\cos(\beta) + \cos(\alpha) + \cos(t_0) - \cos(\alpha + \beta - t_0)),$$
$$y = \frac{1}{2}(\sin(\beta) + \sin(\alpha) + \cos(t_0) - \cos(\alpha + \beta - t_0)). \tag{3.3}$$

Exercise 3.2. Derive equations (3.3). Use CAS assistance if you wish.

This result is just a magic box from which we obtain formulas for the coordinates of the feet of all three perpendiculars:

---

[3] Cramer's rule will be derived and discussed in Chapter 6.

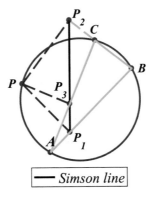

Figure 3.3. An example of a vertical Simson line.

$$x_{P1} = \frac{1}{2}\left(\cos(t_2) + \cos(t_1) + \cos(t_0) - \cos(t_2 + t_1 - t_0)\right),$$

$$y_{P1} = \frac{1}{2}\left(\sin(t_2) + \sin(t_1) + \sin(t_0) - \sin(t_2 + t_1 - t_0)\right),$$

$$x_{P2} = \frac{1}{2}\left(\cos(t_3) + \cos(t_2) + \cos(t_0) - \cos(t_3 + t_2 - t_0)\right),$$

$$y_{P2} = \frac{1}{2}\left(\sin(t_3) + \sin(t_2) + \sin(t_0) - \sin(t_3 + t_2 - t_0)\right),$$

$$x_{P3} = \frac{1}{2}\left(\cos(t_3) + \cos(t_1) + \cos(t_0) - \cos(t_3 + t_1 - t_0)\right),$$

$$y_{P3} = \frac{1}{2}\left(\sin(t_3) + \sin(t_1) + \sin(t_0) - \sin(t_3 + t_1 - t_0)\right).$$

(3.4)

To complete the proof we need to show that the feet of the perpendiculars P_j, $j = 1, 2, 3$, are collinear.

Exercise 3.3. Use formulas (3.4) to verify that the feet of perpendiculars are collinear. Use CAS assistance if you wish.

$\square$

3.2.2 Lab 3: Constructing the Simson line.

Problem formulation. Consider a triangle inscribed in the unit circle C centered at the origin with vertices $\exp(it_j)$, $j = 1, 2, 3$. Let $P(\exp(it_0))$ be a point on C. Make a CAS function that finds the equation of the Simson line and returns a figure with plots of the triangle, the circle C, the three perpendiculars from P to the sides of the triangle or their extensions, and the Simson line.

Suggested plan and directions. Let P_j, $j = 1, 2, 3$ be the feet of perpendiculars dropped from point P to the sides of the given triangle or their extensions.

Part 1. Make a function *orts_feet*(t1, t2, t3, t0) that returns the coordinates of P_j.

Part 2. Make a function *simson_line*(t1, t2, t3, t0) that calls the function *orts_feet*(t1, t2, t3, t0) and returns the coordinates of P_j and the equation of the Simson line.

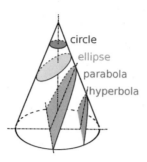

Figure 3.4. Conics as intersections of a cone and some planes. (Figure from [19].)

Part 3. Make a function *simson_plot*(t1, t2, t3, t0) that calls the function *simson_line*(t1, t2, t3, t0) and returns the required figure. Test your function on an example of your choice. Your figure will be similar to Fig. 3.2.

Part 4. (challenge, optional) Design a triangle with a vertical Simson line and use the function made in Part 3 to display your design. Your figure will be similar to Fig. 3.3.

3.3 Conics

Conics, or **conic sections**, are curves obtained by intersecting a double right circular cone with a plane. The three types of conic sections are ellipses, hyperbolas, and parabolas (see Fig.3.4). Special *degenerate* cases occur when the plane passes through the vertex. Such planes intersect the cone at a point, a line, or a pair of intersecting lines.

💡 *Check Your Understanding.* Something is missing in Fig. 3.4 so that we don't see the second branch of the hyperbola. What is missing in this figure?

The general equation of any conic is a second degree polynomial equation

$$A x^2 + B x y + C y^2 + D x + E y + F = 0, \tag{3.5}$$

where the real constants $A, B,$ and C are not all zero. The **discriminant of a conic section** is defined to be $d = B^2 - 4AC$. Equation (3.5) represents an ellipse if $d < 0$ and a hyperbola if $d > 0$. If $d = 0$, the equation defines a parabola or one of the degenerate cases.

3.3.1 Various equations of conics. The standard (canonical) forms for the equations of an ellipse, hyperbola, and parabola are

$$\frac{x^2}{a^2} + \frac{y^2}{b^2} = 1,$$
$$\frac{x^2}{a^2} - \frac{y^2}{b^2} = 1,$$
$$y^2 = 4ax.$$

The parameters a and b in the standard equations of an ellipse and hyperbola are called the **semi-major axis** and **semi-minor axis**, respectively. A parabola defined by its standard equation has the x-axis as its axis of symmetry. More familiar would be the standard equation $x^2 = 4ay$ with the y-axis as the axis of symmetry.

The Pythagorean identity confirms that $x = a\cos(t), y = b\sin(t)$ is a correct parametrization for the standard form of an ellipse. To parametrize a hyperbola defined by its standard form, we use the **hyperbolic sine** and the **hyperbolic cosine** functions:

$$\sinh(t) = \frac{\exp(t) - \exp(-t)}{2},$$
$$\cosh(t) = \frac{\exp(t) + \exp(-t)}{2}.$$

💡 *Check Your Understanding.*

(a) Show that $x = \pm a\cosh(t)$, $y = b\sinh(t)$, $t \in (-\infty, \infty)$, is a correct parametrization of a hyperbola defined by its standard equation.

(b) Find a parametrization of a parabola defined by its standard form.

Remark 3.4. The implicit polynomial and parametric forms of a curve have both advantages and disadvantages. For example, it is easier to decide if a given point lies on a curve using its implicit form, but a parametric formula provides a direct way to generate points on a curve and allows one to think more dynamically about how a point would move along the curve.

Using rotation and translation transformations, the general equation (3.5) can be reduced to the standard form. If $B = 0$, this can be done by the translation $X = x + x_0, Y = y + y_0$ with x_0 and y_0 found by using the **completing the square** technique.

Example 3.5. Consider equation (3.5) with $B = 0$ and $AC \neq 0$. We will find the standard equation of the conic defined by this equation.

Solution. Using the completing the square technique we obtain

$$A\left(x + \frac{D}{2A}\right)^2 - \frac{D^2}{4A} + C\left(y + \frac{E}{2C}\right)^2 - \frac{E^2}{4C} + F = 0. \tag{3.6}$$

Introduce the notation $R = D^2/(4A) + E^2/(4C) - F$ and new variables $X = x + D/(2A), Y = y + E/(2C)$. Notice that the origin of the new coordinate system XY is located at the point $O_1(-D/(2A), -E/(2C))$. If $R \neq 0$, the equation in the new coordinates takes the form

$$\frac{X^2}{R/A} + \frac{Y^2}{R/C} = 1.$$

If $AC < 0$, this equation defines a hyperbola. If $AC > 0$, this equation defines either the empty set or an ellipse.

💡 *Check Your Understanding.*

(a) What is the locus[4] of equation (3.6) if $R = 0$?

---

[4]In geometry, a locus is the set of all points whose location satisfies or is determined by one or more specified conditions.

(b) When does equation (3.6) define the empty set?

Exercise 3.6.

(a) Consider the equation $2x^2 - 3y^2 + 4x + 9y - 31 = 0$. Find the translation transformation that converts the equation to standard form. Use CAS assistance to plot the curve along with the old and new coordinate systems in one figure.

(b) Consider the general equation of a conic with $B = 0$. Make a CAS function *standardize_conic*(**A, C, D, E, F**) that takes coefficients A, C, D, E, F of this equation and returns the standard equation of this curve in the new coordinates. Test your function on the conic given in part (a).

In the example below we will find a rotation that eliminates the product term Bxy from the general equation (3.5) of a conic with $D = E = 0$ and $AC \neq 0$.

Example 3.7. Consider the equation $A x^2 + Bx y + C y^2 + F = 0$.

(a) Show that the change of variables using the rotation through the angle t with $\cot(2t) = (A - C)/B$ eliminates the product term from the given equation.

(b) Let $A = C = 2$, $B = -1$, $F = -1$. Find the angle t of rotation that eliminates the product term Bxy from the equation with these parameters.

(c) Plot the curve defined by the parameters in part (b) in the original rectangular xy-coordinate system. Include the new XY-coordinate system.

Solution.

(a) Let t be the angle of rotation. The old coordinates are related to the new ones by the formulas

$$x = X \cos(t) - Y \sin(t),$$
$$y = X \sin(t) + Y \cos(t).$$

We substitute these expressions into the given equation and find the coefficient of the product $X Y$ to be (**check this!**)

$$-2A \sin(t) \cos(t) + 2B\big(\cos^2(t) - \sin^2(t)\big) + 2C \sin(t) \cos(t).$$

Equating this coefficient to zero and simplifying using the trigonometric identities $\cos^2(t) - \sin^2(t) = \cos(2t)$ and $2 \sin(t) \cos(t) = \sin(2t)$, we obtain

$$(A - C) \sin(2t) = B \cos(2t) \Rightarrow \cot(2t) = \frac{A - C}{B}.$$

(b) We have to find the value of the angle t or values of $\sin(t)$, $\cos(t)$ to implement the required rotation. In general, some trigonometric manipulations are needed to find sine and cosine of an angle t when $\cot(2t)$ is known. But for the parameter values given in this part of the example, these manipulations can be avoided since $\cot(2t) = 0$, which implies that $t = \pm\pi/4$. Choosing $t = \pi/4$, we find the rotation transformation

$$x = (X - Y)\sqrt{2}/2, \; y = (X + Y)\sqrt{2}/2.$$

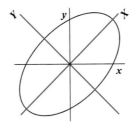

Figure 3.5. Ellipse represented by the standard equation in the new coordinate system XY.

The given equation in the new coordinates then takes the form

$$2\left((X-Y)\frac{\sqrt{2}}{2}\right)^2 - \left((X-Y)\frac{\sqrt{2}}{2}\right)\left((X+Y)\frac{\sqrt{2}}{2}\right) + 2\left((X+Y)\frac{\sqrt{2}}{2}\right)^2 = 1.$$

Simplifying, we obtain

$$\frac{X^2}{1} + \frac{Y^2}{1/3} = 1.$$

(c) See Fig. 3.5

Remark 3.8. How can one find $\sin(t)$ and $\cos(t)$ exactly when $\cot(2t)$ is known? Using the identity $\cot(t) - \tan(t) = 2\cot(2t)$, we can find the value(s) of $\tan(t)$ (**how?**) and then express $\cos(t)$ and $\sin(t)$ in terms of this value using appropriate trig identities.

🔆 ***Check Your Understanding.***

(a) Prove the identity $\cot(t) - \tan(t) = 2\cot(2t)$.

(b) Let $\cot(2t) = c$. Find $\tan(t)$, $\sin(t)$, and $\cos(t)$ as functions of c.

3.3.2 Mini project: Reduction of the equation of a conic to standard form.

Problem formulation. Consider the equation $3x^2 + 6xy - 3y^2 - 1 = 0$. Use an appropriate rotation to transform the equation into standard form. Plot the conic defined by this equation and the old and new coordinate systems in one figure. Your figure will be similar to Fig. 3.5.

Suggested directions. Follow the steps in the previous example to find the angle of the required rotation angle t. If $\sin(t)$ and $\cos(t)$ cannot be evaluated exactly, approximate their values to three decimal places. Use these approximations to derive the equation of the given curve in the new coordinate system.

3.3.3 The geometric definitions of conics. For any non-circular, non-degenerate conic there exists a line called the **directrix** and a point called the **focus**.

Such a conic can be defined geometrically as the locus of points whose distances to the **focus** and **directrix** are in a fixed ratio. This ratio is called the **eccentricity**, which is typically denoted as e. An ellipse can be defined as the locus of points whose distances from the focus and directrix are in a fixed ratio $e < 1$. A hyperbola can be defined as the locus of points whose distances from the focus and directrix are in a fixed ratio $e > 1$. A parabola is the locus of points equidistant from the focus and the directrix.

Example 3.9. Suppose that a line is drawn on a plane without a coordinate system and a point not lying on the line is given. Consider the point and line as the focus and directrix of a parabola. Derive the equation of the parabola in the rectangular coordinates chosen as follows:

- The x-axis is the line perpendicular to the directrix, passing through the focus and directed from the directix to the focus.

- The y-axis is the line passing through the midpoint of the x-axis segment between the focus and the directrix.

Solution. The origin lies on the parabola. (Why?) Denote the distance between the focus and the directrix as $2a$. Then the focus is located at $F(a, 0)$, and the directrix is defined by the equation $x = -a$. By definition, for any point (x, y) on the parabola the following equation holds:
$$\sqrt{(x-a)^2 + y^2} = x + a.$$
Solving for y^2 yields the canonical equation of the parabola $y^2 = 4ax$.

Definition 3.10. An ellipse is the locus of points such that the sum of their distances from two fixed points is a constant. Each fixed point is called a focus.

Let the focus $F(c, 0)$, $c > 0$, and semi-major axis a, $a > c$, of an ellipse be given. It can be shown that the ratio c/a equals the eccentricity. (Show this!) The eccentricity is considered as a measure of deviation of this shape from a circle.

Definition 3.11. A hyperbola is the locus of points such that the absolute value of the difference of their distances from two fixed points is constant. Each fixed point is called a focus.

Similar to that of an ellipse, given the focus $F(c, 0), c > 0$, and semi-major axis a, $a < c$, of a hyperbola, it can be shown that the ratio $e = c/a$ equals the eccentricity of the hyperbola. (Show this!) The eccentricity measures the degree of the opening of the hyperbola's branches. The greater the eccentricity, the larger the opening.

Exercise 3.12.

(a) Given the focus $F(3, 0)$ and the directix $x = 19/3$ of an ellipse, find its standard equation.

(b) Given the focus $F(4, 0)$ and the directrix $x = 2.25$ of a hyperbola, find its standard equation.

Hint: To find the semi-major axes of these conics, equate the eccentricity as the ratio of distances to the focus and directrix to c/a. To find the semi-minor axes, use Definitions 3.10 and 3.11 of an ellipse and hyperbola.

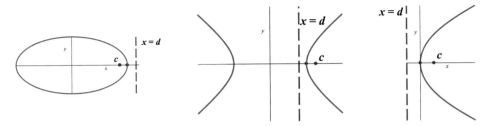

Figure 3.6. Conics defined by standard equations found using the given focus and directrix.

📖 Reflective surfaces in the shape of a **paraboloid of rotation**[5] are used to collect or project energy such as light, sound, or radio waves. Waves that travel parallel to the axis of symmetry of such a paraboloid and strike its concave side are reflected to its focus,[6] regardless of where on the paraboloid the reflection occurs. Conversely, waves that originate from a point source at the focus are reflected into a parallel ("collimated") beam that is parallel to the axis of symmetry.

Exercise 3.13. Make a function **my_conic(c, a)** that takes the x-coordinate $c > 0$ of the focus $F(c, 0)$ and the semi-major axis $a > 0$, and returns the standard equation of the conic (ellipse or hyperbola) defined by the inputs and a figure with the conic, one of the two directrices, and the given focus. The figure will be similar to one of the plots in Fig. 3.6. Test the function on data of your choice.

3.4 Glossary

- **Euler line** – The line passing through the **center of mass, orthocenter**, and **circumcenter** of a triangle.

- **Simson line** – The line passing through the feet of the perpendiculars from an arbitrary point P on the circumcircle of a triangle to the sides of the triangle or their extensions.

- **Uniform scaling** – A linear transformation that enlarges or shrinks an object by a scale factor that is the same in all directions. It is known that a uniform scaling changes distances but does not change angles.

- **General equation of a conic:**

$$A x^2 + 2 B x y + C y^2 + D x + E y + F = 0.$$

The expression $d = B^2 - 4AC$ is called the **discriminant of a conic**. The equation defines an ellipse if $d < 0$ and a hyperbola if $d > 0$. For $d = 0$ the equation defines a parabola or one of the degenerate cases (a line, a pair of intersecting lines, or a point).

---

[5]This is the surface obtained by rotating a parabola about its axis of symmetry.
[6]which coincides with the focus of the parabola that generated the surface

- **Standard equations of conics**:

$$\frac{x^2}{a^2} + \frac{y^2}{b^2} = 1 \text{ (ellipse)}; \quad \frac{x^2}{a^2} - \frac{y^2}{b^2} = 1 \text{ (hyperbola)}; \quad y^2 = 4a\,x \text{ (parabola)}.$$

The parameters a and b in the equations of an ellipse and hyperbola are called the **semi-major axis** and **semi-minor axis**, respectively.

- **Ellipse/hyperbola** – The locus of points such that the sum/absolute value of the difference of their distances from two fixed points called **foci** is constant.

- **Eccentricity of a non-circular, non-degenerate conic** – The ratio e of distances of any point on a conic from a fixed point, called the **focus**, and a fixed line, called **directrix**. For an ellipse $e < 1$ and for a hyperbola $e > 1$. For a parabola $e = 1$.

4

Topics in Elementary Number Theory

Number theory is the field of mathematics devoted primarily to the study of the integers and integer-valued functions. Some typical examples of number theoretic questions are:

(1) Can the sum of two squares of integers be a square?

(2) Is the equation $x^n + y^n = z^n$ solvable for $n > 2$?

(3) Is the set of primes finite or infinite?

(4) How to determine if a given integer is prime or composite?

(5) How to efficiently factor a composite number?

The first two questions are about the existence of positive integer solutions of the equations $x^n + y^n = z^n, n \geq 2$. Certainly, the equation is solvable for $n = 2$. Its solutions are called **Pythagorean triples** (PTs). A PT with coprime integers x and y (that is, x and y do not have any common divisors other than one) is called a **primitive Pythagorean triple** (PPT). Two characterization theorems for PPTs are included in this chapter. The first theorem will be used in Lab 1, where a large number of PPTs will be constructed, visualized, and explored.

Fermat's Last Theorem states that there is no triple of positive integers x, y, z satisfying the equation $x^n + y^n = z^n$ with $n > 2$. Fermat did not provide the proof, and the statement remained just a conjecture for more than three centuries. In 1994, British mathematician Andrew Wiles presented the first successful proof – after 358 years of effort by mathematicians!

The equation in question two is a particular example of what is called a Diophantine equation. A general Diophantine equation in two variables is of the form $P(x, y) = 0$, where $P(x, y)$ is a polynomial with integer coefficients; solutions are sought in the set of integers. It is known that a general algorithm to solve any Diophantine equation does not exist. In this chapter we will introduce only linear Diophantine equations

(LDEs) in two and three variables. An LDE in two unknowns is always solvable under a simple assumption, and we will show a solution method using an elementary example. Lab 2 in this chapter is on solving an applied problem modeled by an LDE in three unknowns.

The third question was answered by Euclid. Euclid's proof of the theorem on the infinitude of primes is simple and beautiful. Just google "Euclid's Proof of the Infinitude of Primes", choose a source page, and enjoy this more than 2000-year-old masterpiece of reasoning.[1]

The last two questions are about primality testing and prime factorization. More than 2000 years ago, Euclid and Eratosthenes developed algorithms for these tasks. In the 17th century, Pierre de Fermat made the crucial step in studying primality by stating a necessary condition for a number to be prime. Euler then proved this statement almost a hundred years later. The result has since become known as "Fermat's little Theorem". Euler, along with other brilliant mathematicians, such as Jacobi and Gauss, deepened Fermat's results. Carl Friedrich Gauss considered primality and factorization as the central problems in number theory.

These are now not just academic questions. Primality and prime factorization have become very important in applications, especially in cryptography. In the 1970s it was barely possible to factor "hard" 20-digit numbers. Presently, the general number field sieve (GNFS) is considered the most efficient classical algorithm known for factoring integers larger than 10^{100}. Modern fast prime testing algorithms have also been developed, but these use sophisticated approaches that cannot be applied directly to integer factorization. An efficient deterministic algorithm for testing primality called AKS, using the first initials of the three co-authors (M. Agrawal, N. Kayal, and N. Saxena), was found in 2002. It was a breakthrough in primality testing and the first fully proven *polynomial-time* test. Many algorithms have been devised to make factorization ever faster. It is a major open problem in number theory to determine how fast integer factorization algorithms can be. Currently, there are no polynomial-time algorithms to factor a large integer N into its prime factors.

In this text, we describe only four classical algorithms to give just a glimpse of intimately related problems of primality testing and integer factorization.[2] Although the algorithms can only be used for primality testing and factoring of integers that are not too large, the insights and ideas in these centuries-old algorithms are still useful today in developing mathematically justified methods for solving these two problems.

There are open problems in number theory stated as conjectures and verified for a huge number of cases using modern computers, but not proven rigorously. Here is a classic example of such a conjecture.

Goldbach's Conjecture. *Every even number $n \geq 4$ is the sum of two primes. For example,* $6 = 3 + 3$, $10 = 3 + 7$, $100 = 3 + 97$.

The conjecture has been tested up to the number $4 \cdot 10^{18}$ (2021), but remains unproven.

Exercise 4.1. Make a CAS function *my_goldbach*(n) that takes an even number $n > 2$ and prints all pairs p_j, q_j of prime numbers p_j, q_j that add up to n.

---

[1] Or see, for example, p. 3 of "The New Book of Prime Number Records" by Paolo Ribenboim.
[2] For further details on primality testing and integer factorization, see [5].

Suggestion. Use a *while* loop with the loop variable starting from $p_1 = 2$ and built-in CAS functions for testing the primality of $q_j = n - p_j$ and finding the prime next to p_j. Manually experiment with small even numbers to understand how to formulate the conditional statement for the loop, that is, how far to go with the primes p_j for the given input n.

📖 In modern times number theory has turned out to become one of the most useful branches of mathematics due to its applications in computer security. For example, algorithms based on number theory are used to protect the transfer of sensitive information, such as credit card numbers when shopping online. Another important practical issue is the necessity to accelerate the execution of numerical calculations by computers. According to computer scientist Donald Knuth, who has been called the "father of the analysis of algorithms," "virtually every theorem in elementary number theory arises in a natural, motivated way in connection with the problem of making computers do high-speed numerical calculations."[3]

4.1 Number of primes and the Riemann Hypothesis

Before diving into algorithms for implementing various tasks related to prime numbers, we introduce an object of the greatest interest in number theory called the prime-counting function. Its exact mathematical form is rather sophisticated, but for large values of inputs, the prime-counting function can be approximated using an elementary function.[4] We will demonstrate using this approximation for some constructive tasks in a very simple exercise at the end of this section.

The prime-counting function, denoted $\pi(x)$, counts the number of primes less than or equal to a given positive number x. It is known that $\pi(x)$ can be approximated by the elementary function $x/\ln(x)$ in the sense that the limit of the ratio of these two functions when x grows without bound equals one. That is, $\lim_{x\to\infty} \ln(x)\cdot\pi(x)/x = 1$. This is the statement of the **Prime Number Theorem**. It is also known that the famous Riemann hypothesis is equivalent to the conjecture that the function $li(x) = \int_2^x \frac{dt}{\ln(t)}$ is an approximation of $\pi(x)$ in the same sense, that is, $\lim_{x\to\infty} \frac{\pi(x)}{li(x)} = 1$.

Remark 4.2. It is known that the function $li(x)$ is not elementary. We are lucky to have CAS assistance and appropriate built-in CAS functions to evaluate and visualize $li(x)$ and $\pi(x)$.

Exercise 4.3. Make a function *count_primes*(n) that returns a figure with plots of $\pi(x)$, $li(x)$, and $x/\ln(x)$ in the range $[3, n]$. Choose a large integer n and use the function to make a figure similar to Fig. 4.1.

🔆 *Check Your Understanding.* Clearly the prime counting function is piecewise constant. Plot it for some small values of x to visualize this. Make sure your graph

---

[3]D. Knuth, Computer science and its relation to mathematics, The American Mathematical Monthly, Vol. 81, No. 4, 1974, p 327.
[4]See a description of the class of elementary functions in the section Glossary.

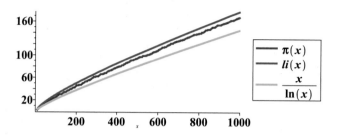

Figure 4.1. The prime-counting function and its asymptotic approximations.

does not include vertical segments at the point where the function is not continuous.[5] Describe in words the points of discontinuity of $\pi(x)$. Is the function $\pi(x)$ continuous from the left or from the right at these points?

We can use the Prime Number Theorem to approximate the probability of randomly choosing a prime number below a certain large N. In fact, there are N possible outcomes of this random experiment and about $\pi(N)$ of them are favorable for our purpose. Thus, our chances are $\pi(N)/N$. Apply this argument in the next exercise.

Exercise 4.4.

(a) Estimate the probability that a randomly chosen number less than 10^{100} is prime.

(b) How many numbers should we choose in order to be practically sure (i.e., with probability approximately equal to 1) that there will be a prime among them? **Hint:** If p is the answer to part (a), the probability of getting a prime choosing m random numbers less than 10^{100} is $m \cdot p$.

4.2 Algorithms from elementary number theory

4.2.1 The Euclidean algorithm revisited. The Euclidean algorithm returns the greatest common divisor $\gcd(a, b)$ of the integers a, b. Without loss of generality assume that $a, b > 0$. Recall from Section 1.3.3 that the algorithm uses Euclidean division to find the quotient and remainder of the integer division of a by b. For example, for the dividend $a = 203$ and the divisor $b = 91$, the quotient is 2 and the remainder is 21. Thus, we write $203 \bmod 91 = 21$. The logic of the algorithm is nicely described by the pseudocode in Chapter 1:

---

[5]For example, SageMath uses the optional variable `VerticalLines=False` and Maple uses the optional variable `discont=True`. In SymPy vertical lines are excluded by default.

Algorithm Euclidean Algorithm

1:	**function** EUCLID(a, b)	▷ The gcd of a and b
2:	$r \leftarrow a \bmod b$	▷ Set r to $a \bmod b$
3:	**while** $r \neq 0$ **do**	▷ Repeat until $r = 0$
4:	$a \leftarrow b$	▷ If condition holds, set a to b
5:	$b \leftarrow r$	▷ Set b to r
6:	$r \leftarrow a \bmod b$	▷ Update r
7:	**return** b	▷ The gcd is b

Notice that on each step in the *while* loop the current pair of numbers (a, b) is replaced with the new one $(b, a \bmod b)$.

Exercise 4.5. Use the pseudocode above to make a CAS function *my_euclid*(**a**, **b**) that takes two positive integers a, b and returns their gcd. Use the built-in modulo function for finding the remainder.

4.2.2 Representing numbers in different bases. The traditional decimal notation that we use in everyday life is called a base-10 number system, meaning that the position of each decimal digit is based on powers of 10. Mathematically, the general base-10 representation of an integer with k digits has the form

$$n = 10^{k-1} \cdot n_{k-1} + 10^{k-2} \cdot n_{k-2} + \cdots + 10^1 \cdot n_1 + 10^0 \cdot n_0, \tag{4.1}$$

where the coefficients $n_0, n_1, \ldots, n_{k-1}$ are integers that range between 0 and 9 inclusive. A brief version of this form is $(n_{k-1}n_{k-2} \ldots n_1 n_0)_{10}$. For example,

$$21387 = 10^4 \cdot 2 + 10^3 \cdot 1 + 10^2 \cdot 3 + 10^1 \cdot 8 + 10^0 \cdot 7 = (21387)_{10}.$$

In general, a base-m number system has the same format as (4.1), but with the number 10 replaced by m:

$$n = m^{k-1} \cdot n_{k-1} + m^{k-2} \cdot n_{k-2} + \cdots + m^1 \cdot n_1 + m^0 \cdot n_0, \tag{4.2}$$

where the coefficients $n_0, n_1, \ldots, n_{k-1}$ are integers that range between 0 and $m - 1$.

The binary number system. The base-2 number system, called the **binary number system**, is crucial in computer science. In fact, all of computer processing and communication is carried out using binary numbers. The reason for this is because computers need a convenient way of expressing whether something is "on" or "off." In particular, the circuits in a computer's processor are made up of billions of transistors that are activated by electronic signals. The digits 0 and 1 (called *bits*), which are the only digits allowed in the binary system, reflect the on and off states of these transistors.

To convert a binary number to a traditional base-10 decimal, we use formula (4.2) with base $m = 2$. Since the only binary digits are 0 and 1, we just add powers of two that correspond to 1's in the binary representation.

Example 4.6. Convert $(10110011)_2$ to base-10.

Solution. By definition of a base-2 number we have $2^7 + 2^5 + 2^4 + 2 + 2^0 = 179$. Thus, $(10110011)_2 = (179)_{10}$.

To convert an integer in base-10 to binary, we repeatedly divide the given integer by 2 until the result is less than 2. At each division step, we note the remainder. The desired binary number is then just the combination of all remainders written in the reverse order.

Example 4.7. Convert 14 to binary.

Solution. $14 = 7 \cdot 2 + 0$, $7 = 3 \cdot 2 + 1$, $3 = 1 \cdot 2 + 1$, $1 = 0 \cdot 2 + 1$. Thus, $(14)_{10} = (1110)_2$.

Exercise 4.8.

(a) Convert $(317)_{10}$ to binary.

(b) Convert $(100110101)_2$ to base-10.

(c) Make a CAS function ***base10_to_base2(n)*** that converts a positive decimal integer n into its binary representation. Test the function on some examples and check the results using an appropriate CAS command.

   Suggestion: Make a list of strings of the remainders found by the CAS modulo function. Then the binary representation of the input can be found using an appropriate concatenation (merging) operation on the elements in the list.

Any natural number N having a binary representation of length b must satisfy the inequality

$$2^{b-1} \le N \le 2^b - 1. \tag{4.3}$$

(**Think why.**) Use this inequality to do the next exercise.

Exercise 4.9.

(a) Prove the formula $b = \lfloor \log_2(N) \rfloor + 1$ for the number of digits b in the binary representation of a natural number N. **Hint:** Apply the logarithm with base 2 (binary logarithm) to all sides of the inequality (4.3). Then apply the floor function and think how to complete the proof.

(b) Use CAS assistance to verify the formula for several natural numbers of your choice.

4.2.3 Primes, primality tests, and prime factorization. A **prime number** is a natural integer greater than one with no positive integer divisors other than one and itself. Two integers are called **mutually prime**, or **coprime**, if the only common divisor of these integers is one. More generally, a set of integers is coprime if the only common divisor of the members of the set is one. The **Fundamental Theorem of Arithmetic**, also called the **Unique Factorization Theorem**, states that every integer greater than one is either a prime number itself or can be represented uniquely as a product of prime numbers, up to the order of factors. The theorem is a typical existence statement; it does not provide a way to constructively factor composite numbers or determine if a number is prime or composite. Note that primality tests and complete factorization problems are closely related. Without a method for determining primality, we have no way of knowing when the complete factorization of a number is done.

The Sieve of Eratosthenes and trial division. The **Sieve of Eratosthenes** is an ancient algorithm that finds all primes less than or equal to a given integer n. It can be used as a primality test for integers that are not too large. Here are the steps of the algorithm:

(1) Make a list of all natural numbers from 2 to n.

(2) Set $p = 2$.

(3) Cross out all numbers divisible by p except for p itself.

(4) Find the first non-crossed number l on the list that is greater than p. If there is no such number, stop. Otherwise, set $p = l$, which is the next prime, and repeat from step 3.

Once you have crossed off all multiples of all primes less than or equal to $\lfloor \sqrt{n} \rfloor$, then anything not crossed out must be prime.

Example 4.10. Implement the Eratosthenes algorithm manually for $n = 50$.

Solution. We made a list of natural numbers from 2 to 50 and consecutively eliminated multiples of primes up to and including the number $\lfloor \sqrt{50} \rfloor = 7$. Here are the outputs after excluding multiples of 2, 3, 5 and 7:
$[2, 3, 5, 7, 9, 11, 13, 15, 17, 19, 21, 23, 25, 27, 29, 31, 33, 35, 37, 39, 41, 43, 45, 47, 49, 50]$
$[2, 3, 5, 7, 11, 13, 17, 19, 23, 25, 29, 31, 35, 37, 41, 43, 47, 49]$
$[2, 3, 5, 7, 11, 13, 17, 19, 23, 29, 31, 37, 41, 43, 47]$.
The last list is the answer.

Exercise 4.11.

(a) Make a CAS function **my_erato**(n) that returns the list of all prime numbers less than or equal to n using the Sieve of Eratosthenes. Test the function on the problem in Example 4.10.

(b) Make a CAS function **list_of_primes**$(\mathbf{k}, \mathbf{n})$ that makes a list of prime numbers between the positive integers $k1$ and $k2$ with $k1 < k2$. Test the function on a simple example.

Suggestion: For part (a), use a *while* loop similar to the one that was manually implemented in Example 4.10. For part (b), make two lists $L1$, $L2$ of primes calling the function created in part (a) with the parameters $k1$, $k2$. Then use an appropriate CAS operation to remove elements of $L1$ from $L2$.

The **trial division** method (also called the **brute force**[6] method) is the simplest primality test. It is a deterministic algorithm that works as follows: Given a number n, all integers less than or equal to $\lfloor \sqrt{n} \rfloor$ are tested using trial division to see if they actually divide the given number. Used repeatedly, trial division works as a factorization algorithm, but is slow and not practical for large n. The next example demonstrates this inefficiency by finding a crude estimate of the time needed for deciding if a large number is prime or composite.

---

[6]Brute forcing is an exhaustive search method which tries all possibilities to reach the solution of a problem.

Example 4.12. Suppose we use the trial division algorithm to decide whether an integer n of 100 decimal digits is prime or composite. If we use only primes below $\sqrt{n}$, it would take approximately $\sqrt{10^{100}}/\ln\left(\sqrt{10^{100}}\right)$ (Prime Number Theorem in action!) trial divisions to determine if n is prime or composite. With a processor executing 10^{11} operations in second, it would take approximately $8.69 \cdot 10^{36}$ seconds, or $2.76 \cdot 10^{29}$ years.

This level of complexity requires us to take into account the properties of primes in order to design more efficient primality tests and factorization methods. Various methods for these two mathematical tasks have been developed over the years. Pierre de Fermat was probably the first who made a major breakthrough in solving these problems.

Theorem 4.13 (Fermat's little theorem). *If p is a prime and a is any integer not divisible by p, then $a^{p-1} - 1$ is divisible by p.*

The conclusion of the theorem can be written as $a^{p-1} \bmod p = 1$. This implies that if the conclusion is not true, then p is not a prime. Thus, there are no false negatives in the sense that any integer that fails the test is composite. However, the theorem cannot be used for identifying primes. There are composite numbers for which the conclusion of Fermat's Little theorem is true for a certain base a. They are called **Fermat pseudoprimes to base** a. For $a = 2$, the smallest pseudoprime is found to be $n = 341$. In fact, we have $341 = 11 \cdot 31$, but $2^{341-1} \bmod 341 = 1$, and so it passes Fermat's test for the base $a = 2$. Things get even worse. There are numbers called *Carmichael numbers* for which the conclusion of Fermat's little theorem is true for all relatively prime a-values, but we will not dive into these deeper waters of the theory. Instead, in the next exercise we put the theorem to work in experimenting with a few pseudoprime numbers.

Exercise 4.14. Consider the bases 2, 3, 4. It is known that two of the numbers in the list [1729, 4371, 2047] are pseudoprimes to two of the given bases, and one is a pseudoprime to one of the bases. Use an appropriate CAS factorization command to show that the numbers in the list are composite. Then use the Fermat's little theorem to match each of the pseudoprimes with the base(s) that allow the number to cheat, that is, satisfy the conclusion of the theorem.

Fermat's factorization method. Below we state and prove a theorem about Fermat's factorization method that provides a procedure for representing an odd integer as the difference of two squares. Using Fermat's basic idea and its analysis, various improvements for accelerating the method have been suggested over the years.

Theorem 4.15. *Any odd integer N can be represented as the difference of integer squares.*

Proof. Let N be an odd number and suppose $N = c \cdot d$. Then we can write

$$N = \frac{(c+d)^2 - (c-d)^2}{4} = \left(\frac{c+d}{2}\right)^2 - \left(\frac{c-d}{2}\right)^2.$$

Set $a = (c+d)/2$, $b = (c-d)/2$. Since both c and d are odd (**why?**), $c+d$ and $c-d$ are even. Therefore a and b are integers and $N = a^2 - b^2$. □

The basic algorithm for Fermat's factorization method consecutively traverses through integers $a \geq \lceil \sqrt{N} \rceil$ until the difference $a^2 - N$ becomes a perfect square. Then b is set to $\sqrt{a^2 - N}$ and the factors c, d of N are found as $c = a + b, d = a - b$. Note that if N is prime, either c or d equals one, and this is why Fermat's method can be used for primality testing.

Example 4.16. Factor $N = 11703$ manually using Fermat's factorization algorithm.

Solution. We have $\sqrt{N} \approx 108.18$, so $a_0 = 109$. Let us organize our manual implementation of Fermat's algorithm in a table:

k	0	1	2	3
a_k	109	110	111	112
$\sqrt{a_k^2 - N}$	≈ 13.34	≈ 19.92	≈ 24.86	29

Thus $a = 112, b = 29$.
Answer: Factors of N are $c = a + b = 141$, $d = a - b = 83$. Check: $141 \cdot 83 = 11703$.

Exercise 4.17. Let $a_0 = \lceil \sqrt{N} \rceil$ and $b_0^2 = a_0^2 - N$. There is a nice pattern in computing $b_1^2 = a_1^2 - N$ with a_1 replaced by $a_0 + 1, b_2^2$ with a_2 replaced by $a_1 + 1$, etc. Using this pattern, there is no need to square a_k for $k \geq 1$. Find the pattern (formula) and use it in the next exercise.

The next exercise employs Fermat's factorization algorithm for both factoring and primality testing. It is helpful (but not required) to write pseudocode for the algorithm before furnishing it with code.

Exercise 4.18. Make a CAS function **fermat_factors(n)** that takes an odd number n, implements Fermat's factorization algorithm using the trick found in the previous exercise, and returns the string "The input number is prime." if one of the factors equals one. Otherwise, the function returns two factors found by the algorithm. Run the function on simple examples of your choice and check the results with an appropriate CAS factoring command.

Suggestion: Use a *while* loop with a condition that tests if the current b-value is a perfect square. Use an appropriate CAS function that implements the test.[7] Use an *else-if* conditional to encode two possible output options.

📖 The original classical methods presented in this subsection are impractical for modern applications where there is a need to find huge primes or to factor huge numbers. One such important application is public-key cryptography, or asymmetric cryptography. It is an encryption scheme that uses a pair of mathematically related *keys* called a *public key* and a *private key*. The public key is used to encrypt a message and the private key is used to decrypt it. The most widely recognized and used modern encryption method is a cryptosystem by Rivest, Shamir, and Adleman called the RSA algorithm. At the encryption stage a user needs two large prime numbers, and the

---

[7] For example, in SymPy the function name (identifier) is *is_square*.

public key is defined to be the product of these numbers. To decrypt a message, an adversary person who has only the public key needs factorization techniques. There are numerous detailed descriptions of this cryptosystem at various levels of accessibility that the reader can find on the web.

4.3 Pythagorean triples

A triple (x, y, z) of positive integers is a **primitive Pythagorean triple** (PPT) if $x^2 + y^2 = z^2$ and $\gcd(x, y) = 1$. The next theorem on a characterization of PPTs is based on Euclid's formula for generating Pythagorean triples.

Theorem 4.19 (PPT characterization, version 1). *Given coprime integers m, n of different parity[8] with n > m > 0, the triple (a, b, c) with*

$$a = n^2 - m^2, \ b = 2mn, \ c = m^2 + n^2 \qquad (4.4)$$

is a PPT. Conversely, any PPT (a, b, c) is defined by equations (4.4) with some coprime integers n > m > 0 not both odd.

⚡ *Check Your Understanding.*

(a) Show that if m, n, are odd coprime numbers, then a, b, c defined by equations (4.4) are even numbers. That is, the triple (a, b, c) is a PT but not a PPT.

(b) Use the formulas in (4.4) to find all PPTs with $b = 48$.

(c) Does a PPT with $b = 30$ exist? If yes, find the PPT. If no, explain why.

Exercise 4.20. Show that for any Pythagorean triangle it is possible to find the unique triple of positive integers such that if one places the integers at the vertices, the sum of the numbers at the ends of each side will be equal to the length of the side. (A right triangle is called Pythagorean if its side lengths form a Pythagorean triple.)

We now present an alternative characterization of PPTs.

Theorem 4.21 (PPT characterization, version 2). *Given any odd coprime integers s, t with s > t ≥ 1, the triple (x, y, z) with*

$$x = st, \ y = (s^2 - t^2)/2, \ z = (s^2 + t^2)/2, \qquad (4.5)$$

is a PPT. Conversely, any PPT (x, y, z) is defined by equations (4.5) with some odd coprime integers s, t, s > t ≥ 1.

For example, choosing $s = 3$, $t = 1$, we obtain the PPT $(3, 4, 5)$. The PPT defined by $s = 7$ and $t = 5$ is $(35, 12, 37)$. It follows from the theorem that for any PPT one of the "legs" x, y is odd and the other is even. (**Show this!**)

⚡ *Check Your Understanding.* Use the formulas in (4.5) to find all PPTs with $x = 45$.

⚡ *Check Your Understanding.* Show that for any PPT the even "leg" is divisible by four.

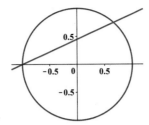

Figure 4.2. Any line $y = t(x + 1)$ crosses the unit circle at the point defined by equations (4.6).

The formulas (4.4) can be derived using geometric considerations as follows. Notice that a triple (a, b, c) of positive integers is Pythagorean if and only if $(a/c)^2 + (b/c)^2 = 1$, that is, the point (x, y) with rational coordinates $x = a/c, y = b/c$ lies on the unit circle centered at the origin. This observation suggests a characterization of PPTs using a rational parametrization[9] of the unit circle centered at the origin. The parametrization

$$x(t) = \frac{1 - t^2}{1 + t^2}, \quad y(t) = \frac{2t}{1 + t^2}, \quad t \in (-\infty, \infty), \tag{4.6}$$

will be derived below.

⚘ Check Your Understanding.

(a) Let $n, m \in \mathbb{N}$, $m < n$, have different parity. Show that if $t = m/n$ in (4.6), we obtain the formulas in (4.4).

(b) When t varies from $-\infty$ to ∞, the point defined by (4.6) traverses the unit circle counterclockwise passing through all points of the circle except the point $(-1, 0)$. Explain why this point is not covered by the rational parametrization.

Trigonometric derivation of a rational parametrization of the unit circle. Recall that the unit circle centered at the origin can be defined parametrically as

$$x = \cos(s), \ y = \sin(s), \ s \in [0, 2\pi).$$

Divide both sides of the trigonometric identities $\cos(s) = \cos^2(s/2) - \sin^2(s/2)$ and $\sin(s) = 2\sin(s/2)\cos(s/2)$ by $\sin^2(s/2) + \cos^2(s/2)$. Next, dividing the numerators and denominators of the obtained right-hand sides by $\cos^2(s/2)$ yields the formulas

$$\cos(s) = \frac{1 - \tan^2(s/2)}{1 + \tan^2(s/2)}, \quad \sin(s) = \frac{2\tan(s/2)}{1 + \tan^2(s/2)}. \tag{4.7}$$

Finally, setting $t = \tan(s/2)$ gives us the rational parametrization of the circle defined by equations (4.6).

---

[8] That is, one is odd and the other is even.

[9] A rational parametrization of a plane curve has the form $x = x(t), y = y(t)$, where $x(t) = p(t)/r(t)$, $y(t) = q(t)/r(t)$, and p, q, r are polynomials.

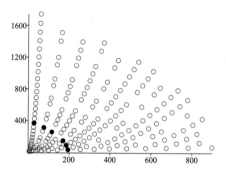

Figure 4.3. Legs of PPTs with "odd" legs on the horizontal axis and
"even" legs on the vertical axis.

Remark 4.22. A rational parameterization of the unit circle can be easily derived geometrically as well. Consider the unit circle in Fig 4.2 centered at the origin and
crossed by a non-vertical line with slope t passing through the point $(-1, 0)$. Substituting $y = t(x + 1)$ into the equation $x^2 + y^2 = 1$, we obtain a quadratic equation for x
with the solutions $x = -1$ and $x(t) = \frac{1-t^2}{1+t^2}$. The latter solution is the first equation in
(4.6). To complete the derivation, substitute $x(t)$ into the implicit equation of the circle
and find $y(t)$.

We complete this section with a simple computational exercise on making two
functions constructing PPTs using formulas (4.4) and (4.5).

Exercise 4.23.

(a) Make a CAS function ***ppt_v2*(s, t)** that takes two odd coprime integers s, t, with
$s > t \geq 1$, and returns the corresponding PPT. Include a verification of mutual
primality. Verifying the inequality condition is not required.

(b) Make a CAS function ***ppt_v1*(r)** that takes a rational number $r = m/n$ and returns
the corresponding PPT or prints "Both m and n are odd."

4.4 Lab 4: Plotting legs of primitive Pythagorean triples

Problem formulation. Consider the following two-way infinite matrix

$$T = \begin{bmatrix} 1/2 & 2/3 & 3/4 & \cdots & n/(n+1) & \cdots \\ 1/3 & 2/4 & 3/5 & \cdots & n/(n+2) & \cdots \\ 1/4 & 2/5 & 3/6 & \cdots & n/(n+3) & \cdots \\ \vdots & \vdots & \vdots & \ddots & \vdots & \ddots \end{bmatrix}.$$

Make a CAS function ***ppt_legs*(N)** that takes a natural number N and selects the following elements of the matrix T:

$$(1/2, \ 1/3, \ 2/3, \ 1/4, \ 2/4, \ 3/4, \ 1/5, \ldots, 1/N, \ 2/N, \ \ldots, \ (N-1)/N).$$

Then the function makes a list L of pairs (m, n), where m is the numerator and n the
denominator of the selected fractions, and returns the list S of PPT legs (a, b) constructed from appropriate elements of L according to the formulas (4.4). Plot the points
$(a, b) \in S$ and answer the questions stated in Parts 4 and 5 of the plan below.

Suggested plan and directions. The solution is more readable if we split the problem into subproblems and make certain help functions.

Part 1. Make a CAS function *fractions_to_pairs*(N) that takes fractions m/n from the first N minor diagonals of the matrix T and returns the list L with distinct pairs $[m, n]$.

Directions: Use nested *for* loops to make a list of all needed fractions. The external loop variable, say i, varies from 0 to $N - 1$. For each i, the range of the internal loop parameter, say j, depends on i. To determine the range of j, find this dependence. Notice that some fractions from the matrix T, for instance 2/3 and 6/9, become identical since the second fraction will be automatically converted into lowest terms. To exclude corresponding equal pairs from the list L you can use the following trick: convert L into a set and then back to a list. The trick works because while a list can have identical elements, a set cannot.

Part 2. Make a CAS function *ppt_legs*(L) that takes a list L of pairs of numbers $[m, n]$ and returns a list S of coprime pairs of integers $[a, b]$ with $a = n^2 - m^2$, $b = 2mn$.
Hint: Think how to encode the condition that a, b are coprime.

Part 3. Make a CAS function *plot_ppt*(N) that calls the function made in Part 2 and returns a plot of the points in the list S. Run this function with $N = 25$. Your plot will be similar to the one in Fig. 4.3. (Six points are made black in the figure just for the next part of the lab.)

Part 4. There are patterns in Fig. 4.3. In particular, it looks like the black points are located along a certain arc and that there is a missing point in this sequence. The points have coordinates

$$(195, 28), (187, 84), (171, 140), \; missing \; point, (115, 252), (75, 308), (27, 364).$$

Determine the pattern of x- and y-coordinate changes in this sequence and find the coordinates of the missing point. Explain why the point is missing.

Part 5. (Challenge) It looks like the points in Fig. 4.3 are located on two types of curves: one starting from the x-axis and going in the northeast direction and the other starting from the x-axis and going in the northwest direction. Find the equations of the curves for each type.

Hint: Consider a set V of points in the xy-plane with coprime coordinates $x = n$, $y = m < n$ such that n is even and m is odd. If $z = x + iy$ then $z^2 = (x^2 - y^2) + i \cdot 2xy$. Thus, the legs of PPTs are images of points in V under the map $z \mapsto z^2$. Use this map to derive the equations of the required curves.

Remark 4.24. Fig. 4.3 suggests that there could be patterns in distributions of points along the lines described in Part 5 of the lab. The reader might want to use codes created for the lab or their modifications to explore if this is true. (We do not know the answer.)

4.5 Linear Diophantine equations in two variables

Let a, b, n be integers. A solution to a **linear Diophantine equation** in two variables (2D LDE)

$$ax + by = n \tag{4.8}$$

is any pair of integers $x, y \in \mathbb{Z}$ that satisfy the equation.

Theorem 4.25 (The Existence Criterion). *A 2D LDE* (4.8) *has a solution if and only if* $\gcd(a, b)$ *divides* n.[10]

Proving that the existence of a solution implies that $\gcd(a, b)$ divides n is trivial. Proving the existence of a solution provided $\gcd(a, b) \mid n$ is more challenging. (**Try it!**)

General solution to 2D LDE. Let (x_0, y_0) be a particular solution to the equation (4.8), that is, $ax_0 + by_0 = n$. It follows that the modified equation $a(b \cdot k + x_0) + b(-a \cdot k + y_0) = n$ is true for all $k \in \mathbb{Z}$. The set $\{(x, y) : x = x_0 + bk, y = y_0 - ak, k \in \mathbb{Z}\}$ is the **general solution of the 2D LDE** (4.8).

Remark 4.26. Notice that if (x_0, y_0) solves the equation $ax + by = 1$, then $(n \cdot x_0, n \cdot y_0)$ is a solution to equation (4.8). That is why a CAS command that solves a 2D LDE typically takes only parameters a, b and returns a solution to the LDE with right-hand side equal to one.

There is a simple algorithm for solving 2D LDEs when the equation is solvable. The next example shows the steps of the algorithm.

Example 4.27. Find all solutions to the LDE $18x + 7y = 5$.

Solution. Since $18 = 7 \cdot 2 + 4$, we can rewrite the given LDE as a "smaller" one in two new variables:

$$(7 \cdot 2 + 4)x + 7y = 5 \Rightarrow 7 \cdot (2x + y) + 4x = 5 \Rightarrow 7u_1 + 4u_2 = 5,$$

where $u_1 = 2x + y, u_2 = x$. Repeat this pattern until the coefficient of the second new variable equals one. Since $7 = 4 \cdot 1 + 3$, we have

$$7u_1 + 4u_2 = 5 \Rightarrow 4 \cdot (u_1 + u_2) + 3u_1 = 5 \Rightarrow 4u_3 + 3u_4 = 5,$$

where $u_3 = u_1 + u_2, u_4 = u_1$.
Finally,

$$4u_3 + 3u_4 = 5 \Rightarrow 3 \cdot (u_3 + u_4) + u_3 = 5.$$

A particular solution to this LDE is $u_3 = 5$, $u_4 = -u_3 = -5$. Now we move backwards to find the original unknowns:

$$-5 = u_4 = u_1 = 2x + y; \quad 5 = u_3 = u_1 + u_2 = 3x + y.$$

Solving the system of two linear equations $2x + y = -5, 3x + y = 5$ we find a particular solution to the original LDE: $x = 10$, $y = -25$.
Check: $18 \cdot 10 + 7 \cdot (-25) = 5$.
Answer: The general solution of the given LDE is $\{(x, y) : x = 10 - 7k, y = -25 + 18k, k \in \mathbb{Z}\}$.

☼ **Check Your Understanding.** Solve the LDE in Example 4.27 in two steps. First, solve the LDE $18x + 7y = 1$. Then construct the general solution to the original LDE $18x + 7y = 5$.

---

[10]In symbolic notation this condition is written as $\gcd(a, b) \mid n$.

Finding nonnegative solutions. In applications, the unknowns often stand for real-life quantities and only nonnegative solutions are sought. If (x_0, y_0) is any particular solution of equation (4.8), nonnegative solutions exist if and only if the system of inequalities

$$b k + x_0 \geq 0, \quad -a k + y_0 \geq 0$$

is solvable for some $k \in \mathbb{Z}$.

$\boxed{\heartsuit}$ **Check Your Understanding.** Show that the LDE in Example 6 has no nonnegative solutions.

A direct approach can be used to find nonnegative solutions of a 2D LDE. Here is how it works. Consider the LDE (4.8) with $a, b, n > 0$ and $a < b$. The idea is to repeatedly reduce n by the value of a and check if the reduced value is divisible by b. If this is true, a corresponding nonnegative solution is defined by the number of subtracted a's and the quotient of the division of the reduced right-hand side by b.

Example 4.28. Find all nonnegative solutions to the LDE $3x + 5y = 23$.

Solution. We have $a = 3, b = 5, n = 23$. The steps of the solution are shown in the table below. The second row is a sequence of positive integers defined as $s_{k+1} = s_k - 3, s_0 = n = 23$.

k	0	1	2	3	4	5	6	7
s_k	23	20	17	14	11	8	5	2
$s_k \bmod 5 = 0$?	no	yes	no	no	no	no	yes	no

Columns one and six with "yes" in the last row yield two nonnegative solutions: $x = 1, y = 20/5 = 4$ and $x = 6, y = 5/5 = 1$.

Note that the number of steps in this example is 8, which is one more than the quotient of n divided by a. On each step k we subtract $a \cdot k$ from n, and if the difference s_k is divisible by b, we obtain the nonnegative solution $x = k, y = s_k/b$. The algorithm can be implemented as a CAS function using a *for* loop and a conditional *if* statement. This will be done in part (b) of the next exercise.

Exercise 4.29.

(a) At a shooting competition participants have two types of targets. If you hit a red target you are awarded 7 points; if you hit a blue target you are awarded 4 points. The winner is a participant who accumulates exactly 50 points with the least number of shots. What is the least possible number of shots to accumulate 50 points?

(b) Make a CAS function **lde2_nonneg_sols(a, b, n)** that implements the direct approach to finding nonnegative solutions to the 2D LDE $a x + b y = n$. Assume that $a < b$. Test your function on the LDE in Example 4.28.

$\boxed{\square}$ The existence of nonnegative solutions to LDEs with n unknowns $\sum\limits_{j=1}^{n} a_j x_j = N$ was studied by mathematicians since the mid-1800s. The problem is also known as the coin problem or money exchange problem. It was found that for each n and sufficiently large N nonnegative solutions exist. The largest number N for which there are no nonnegative solutions, typically denoted $g(a_1, a_2, \ldots, a_n)$, is called the **Frobenius number.** For $n = 2$ the formula for the Frobenius number, $g(a_1, a_2) = a_1 a_2 - a_1 - a_2$, was found at the end of 1800s. For $n \geq 3$, no explicit formula for the Frobenius number is currently known.

4.6 Lab 5: Industrial application of an LDE in three variables

This is an educational version of a real industrial problem solved for a Connecticut manufacturing company.

Problem formulation. An assembly[11] is made of three types of segments: type 1 of length a, type 2 of length b, and type 3 of length c, where a, b, c are pairwise prime integers. We can assume without loss of generality that $a < b < c$. The total length of the assembly L, which we assume is a positive integer, is significantly larger than the length of any segment type. The price of one segment of type j is $p_j, j = 1, 2, 3.$[12] Make a function that takes the total assembly length L, lengths a, b, c, and prices p_j, and returns the number of segments of each type needed to construct the assembly at minimum cost. The function should also return the minimum cost.

Mathematical model. Let $x_j \in \mathbb{N}$ be the number of segments of type $j, j = 1, 2, 3$. The problem can be stated as follows. Minimize

$$f(x_1, x_2, x_3) = p_1 x_1 + p_2 x_2 + p_3 x_3$$

subject to the constraint

$$a x_1 + b x_2 + c x_3 = L, \; x_j \in \mathbb{N}^0.$$

In other words, we want to find a nonnegative integer solution of the LDE $a x_1 + b x_2 + c x_3 = L$ that minimizes the objective function f.

Remark 4.30. This is an example of a so called **integer linear programming** (ILP) problem.

Suggested plan and directions. Split the problem into two parts:

Part 1. Find the feasible set of the ILP. This is a collection of triples $[x_1, x_2, x_3]$ of nonnegative integers that satisfy the constraint of the problem. To accomplish that, make a function ***lde3_nonneg_sols*(a, b, c, L)** that returns a list S of all nonnegative triples $[x_1, x_2, x_3]$ such that for each triple

- $x_3 \in [0, \lfloor L/c \rfloor]$.
- The pair (x_1, x_2) is a nonnegative solution to the 2D LDE $a x_1 + b x_2 = d$ with $d = L - c x_3$.

For solving the 2D LDE $a x_1 + b x_2 = d$, use the function ***lde2_nonneg_sols*** made in part (b) of Exercise 4.29.

---

[11] In Mechanical Engineering an assembly is a group of machine parts forming a self-contained, independently mounted unit.

[12] We also assume that the prices are integers, expressing them in cents if necessary.

Part 2. Make a function *opti_sol*(**lengths**, **price**, **L**) that takes a list *lengths* $= [a, b, c]$ of lengths of three segment types, a list *price* $= [p_1, p_2, p_3]$ of prices for one segment of each type, and the total assembly length L, and does the following:

- Uses the function *lde3_nonneg_sols*(**a**, **b**, **c**, **L**) made in Part 1 to construct the list S of feasible solutions.
- Calculates the cost function for each element $[x_1, x_2, x_3]$ in S and returns the minimal cost and the optimal solution(s).

Test the function on the problem with parameters

$$L = 45 \; a = 3, \; b = 7, \; c = 16, \; p_1 = \$3.75, \; p_2 = \$4.99, \; p_3 = \$5.50.$$

Answer to the test problem: The optimal solution is $[2, 1, 2]$ with cost $\$23.49$.

Exercise 4.31. Use the functions made in the lab to solve the problem with the same parameters as in the test problem but with total length $L = 114$. Write your answer in the context of the industrial problem.

🔲 ILP models have applications in both production planning and various other fields where variables of interest are restricted to be integers. Over the years of developing methods for solving ILPs, numerous algorithms have been designed and analyzed.

4.7 Glossary

- **Prime number** – A positive integer greater than one with no positive integer divisors other than one and itself.

- **Elementary function** – A function of a single variable that is defined as taking sums, products, and compositions of finitely many arithmetic operations and compositions of power functions, exponential functions, trigonometric functions, and their inverses.

- **Coprime set** – A set of integers which do not have common divisors other than one.

- **Fundamental Theorem of Arithmetic**: Every integer greater than one is either a prime number itself or can be represented uniquely as a product of prime numbers, up to the order of factors.

- **Modulo operation** – Finds the remainder after Euclidean division of one integer by another integer. The remainder is called the **modulus** of the operation.

- **Pythagorean triple** – A triple of integers (x, y, z) such that $x^2 + y^2 = z^2$. If x and y are coprime, the Pythagorean triple is called **primitive**.

- **Diophantine equation** in two variables – An equation of the form $P(x, y) = 0$, where $P(x, y)$ is a polynomial with integer coefficients; solutions are sought in the set of integers. An equation defined by a polynomial of degree one, $ax + by = c$, is called a linear Diophantine equation (abbreviated in this text as 2D LDE).

- **General solution of a 2D LDE** $ax + by = c$ – A set $\{x = x_0 + b\,k, y = y_0 - a\,k, k \in \mathbb{Z}\}$, with (x_0, y_0) being a particular solution.

- **Frobenius number** for a linear Diophantine equation $\sum_{k=1}^{n} a_k x_k = N$ – The largest number $g(a_1, a_2, ..., a_n)$ in the right-hand side of the equation for which there are no nonnegative solutions.

5

Topics in Algebra: Solving Univariate Algebraic Equations

Algebra is concerned with algebraic operations defined on sets of symbols. These sets equipped with certain operations satisfying some axioms are called algebraic structures. Specific types of algebraic structures, such as groups, rings, and fields, are studied in abstract algebra courses. This level of abstraction is beyond the scope of this chapter, although some realizations of these structures are certainly familiar to readers. In fact, rotations of a plane with the composition operation form a commutative group. Integers with addition and multiplication operations form a ring. Rational numbers and real numbers with the operations of addition and multiplication are examples of fields.

Polynomial (algebraic) equations and systems of such equations are among the main objects of study in classical algebra. The canonical form for a univariate polynomial of degree n is

$$p(x) = a_n x^n + a_{n-1} x^{n-1} + \cdots + a_1 x + a_0. \tag{5.1}$$

Here, the **leading coefficient** a_n is assumed to be nonzero. Solutions to the polynomial equation $p(x) = 0$ are called **roots** or **zeros** of the polynomial. The problem of finding zeros of polynomials arises in many theoretical fields, including number theory, computer science, and physics, as well as in practical problems in various sciences.[1] Roots of a polynomial of degree less than five can be found in terms of algebraic operations on the polynomial coefficients. A familiar example is the formula for zeros of a quadratic polynomial. It was a Norwegian mathematician, Niels Henrik Abel, who established a theorem that there is no algebraic solution – that is, solution in radicals – to general polynomial equations of degree five or higher with arbitrary complex coefficients. The theorem was superseded by the remarkable results of a French mathematician, Evariste Galois, who developed the solvability theory of polynomial equations.

Typical questions of interest in studying polynomial equations and systems are

---

[1] In solving a real engineering problem similar to the one in Lab 9 in Chapter 8, the exact formula for finding roots of a cubic polynomial was used.

- existence of solutions that belong to specific number sets

- number of solutions that belong to specific number sets

- methods for finding roots of polynomial equations

We will not discuss the remarkable, but cumbersome Cardano's formula for finding roots of cubic polynomials in this chapter. Using the classic Rational Root Theorem, we first introduce a technique for finding rational roots of polynomials with integer coefficients. We will then describe an original geometric technique for counting the number of real solutions of a cubic equation based on the presentation in [**15**]. Next, using the trigonometric form of complex numbers, we will derive the formula for all solutions of the equation $w^n = z$. For $z = 1$ the solutions are the so called **roots of unity**, which have important applications in number theory and signal processing. The lab in this chapter involves constructing and implementing an algorithm for solving cubic equations. The algorithm is based on a clever substitution that transforms a cubic equation into a quadratic one in a new variable. The substitution was invented by Francois Viète, a French mathematician who introduced the first systematic algebraic notation along with other contributions to mathematics. This chapter concludes with a constructive theorem about decomposing any nonnegative polynomial with real coefficients into a sum of squares of two polynomials with real coefficients.

In the next chapter we will explore simple systems of two equations of degree one or two in two variables, both geometrically and algebraically. When roots of a polynomial or, more generally, solutions to a non-polynomial equation $f(x) = 0$ cannot be found exactly, numerical root-finding algorithms are used to approximate solutions. These algorithms employ tools of calculus that will be introduced later in the text.

5.1 Roots of univariate polynomials

In this section we present two algorithms for closed-form solutions of two kinds of polynomial equations. One is for finding rational roots of a polynomial; the other for finding all complex solutions to the equation $w^n = z$ with a given complex number z. We then describe a surprising geometric method for counting the number of real solutions of a cubic polynomial.

The existence of solutions to a polynomial equation depends on the set of numbers in which the solutions are sought. For example, we know that the LDE $15x + 3y = 20$ does not have integer solutions. (**Why?**) However, there are infinitely many real solutions of this equation. These are pairs of numbers $(x, (20 - 15x)/3)$ with $x \in \mathbb{R}$. Similarly, the equation $x^2 + 1 = 0$ does not have any real solutions but has two complex ones, $x = \pm i$.

5.1.1 Number of complex roots.

Definition 5.1 (Multiplicity of a root of a polynomial). If a complex number a is a root of both a polynomial (5.1) and its derivatives p', p'', ..., p^{k-1} such that $p^k(a) \neq 0$, we say that the multiplicity of the root a is k.

Claim. Every univariate polynomial of degree n with complex coefficients has, including multiplicities, exactly n complex roots.

This claim hinges on two facts.

Theorem 5.2 (The Fundamental Theorem of Algebra). *Every non-constant univariate polynomial with complex coefficients has at least one complex root.*[2]

Theorem 5.3 (The Factorization theorem). *If $x = a$ is a root of a polynomial p of degree n, then $p(x) = (x - a)q(x)$, where q is a polynomial of degree $n - 1$.*

In the case of higher-order polynomials, each root can be used to factor the polynomial, thereby simplifying the problem of finding further roots. This involves the **polynomial long division** algorithm that uses Euclidean division of polynomials in a way similar to that of the Euclidean division algorithm for integers. Let $s(x)$ be a polynomial of degree n. Applying Euclidean division to the dividend $s(x)$ and divisor $x - a$ yields

$$s(x) = q(x)(x - a) + r(x),$$

where $q(x)$ is the first quotient and $r(x)$ the first remainder. Then Euclidean division by $x - a$ is executed for the dividend $q(x)$. The process continues until a constant remainder is obtained. The quotient in each step can be found by **dividing the leading term of the polynomial s by x.**

Example 5.4 (Polynomial long division). Implement the polynomial long division algorithm for the dividend $p(x) = x^3 - 6x^2 + 11x - 6$ and divisor $x + 1$.

Solution. Here are the steps of the algorithm:

- Quotient 1: $q_1(x) = x^3/x = x^2$;

  Remainder 1: $r_1(x) = p(x) - x^2(x + 1) = -7x^2 + 11x - 6$.

- Quotient 2: $q_2(x) = -7x^2/x = -7x$;

  Remainder 2: $r_2(x) = r_1 - (-7x)(x + 1) = 18x - 6$.

- Quotient 3: $q_3(x) = 18x/x = 18$;

  Remainder 3: $r_3(x) = r_2 - 18(x + 1) = -24$.

Combining these steps, we can write

$$\frac{x^3 - 6x^2 + 11x - 6}{x + 1} = x^2 + \frac{-7x^2 + 11x - 6}{x + 1}$$
$$= x^2 + \left(-7x + \frac{18x - 6}{x + 1}\right)$$
$$= x^2 - 7x + 18 - \frac{24}{x + 1}.$$

Answer: $p(x)/(x + 1) = x^2 - 7x + 18 - 24/(x + 1)$.

The next theorem is useful for both evaluating polynomials and finding their roots.

Theorem 5.5 (The Remainder Theorem). *The remainder of dividing a polynomial $p(x)$ by $x - a$ is equal to $p(a)$.*

---

[2]Recall that a real root is considered as a complex number with zero imaginary part.

Proof. Using long division, divide $p(x)$ by $x - a$ to obtain $p(x) = (x - a)q(x) + r$, where r is a complex number. Setting $x = a$ in this identity completes the proof.   □

Now the Factorization Theorem just becomes a corollary of the Remainder Theorem. In fact, if $x = a$ is a root of the polynomial p, then $r = p(a) = 0$ and we obtain the required factorization. Repeated application of the Fundamental Theorem of Algebra and the Factorization Theorem proves the claim about the total number of complex roots of a polynomial of degree n.

⌕ **Check Your Understanding.** Show that if $x = a$ is a root of multiplicity k of a polynomial p, then $p(x) = (x - a)^k q(x)$, where q is a polynomial of degree k less than the degree of p.

Synthetic division is a shorthand version of long division. It can be encoded using a loop in the coefficients of the dividend. Below is pseudocode for this procedure. The input is a list A of coefficients of a polynomial of degree n. The procedure returns a list B of length $n + 1$. The first n elements of B are the coefficients of the quotient, which is a polynomial of degree $n - 1$. The last element is the remainder.

Algorithm Synthetic division

1: **function** SYNDIV(A,c)   ▷ Division of polynomial with coefficients A by $x - c$
2:     $B \leftarrow [\,]$ ▷ B is set to the empty list
3:     $val \leftarrow c$ ▷ val is set to c
4:     $B[0] \leftarrow A[0]$ ▷ $B[0]$ is set to the leading coefficient of the dividend
5:     **for** $i \leftarrow 1, n$ **do**   ▷ The index varies from 1 to n
6:         $B[i] \leftarrow A[i] + val \cdot B[i - 1]$ ▷ The next element of B
7:     **return** B

Exercise 5.6. Make a CAS function ***synthetic_div*(A, c)** that implements the synthetic division algorithm. It takes a list A of coefficients of a polynomial p and a complex number c and returns a list B with the coefficients of the quotient and remainder of the division of p by $x - c$.

5.1.2 Finding rational roots of an integer polynomial. In this section we consider **integer polynomials**, which are polynomials with integer coefficients. There is a classic theorem for finding rational roots of integer polynomials called the Rational Root Theorem. It states that if such a polynomial has a rational root m/n written in lowest terms, then m divides the constant term a_0 and n divides the leading coefficient. The theorem suggests a simple brute-force procedure for finding rational roots of integer polynomials. The next example demonstrates this procedure.

Example 5.7. Find all rational roots of the polynomial $p(x) = 3x^3 + 8x^2 - 5x - 6$ using the Rational Root Theorem.

Solution. The divisors of the leading coefficient are $L\_denom = \{\pm 1, \pm 3\}$; the divisors of the constant term are $L\_num = \{\pm 1, \pm 2, \pm 3, \pm 6\}$. The list of candidates for the rational roots of the polynomial p includes all numbers in the form m/n with $m \in L\_num$ and $n \in L\_denom$.

To construct this list for the polynomial p we have to join the list L_num and the list of genuine fractions $m/3$ with $m \in L\_num$. This yields the list of all possible rational roots

$$L = \{\pm 1,\ \pm 2,\ \pm 3,\ \pm 6,\ \pm 1/3,\ \pm 2/3\}.$$

To find all rational roots we evaluate the polynomial at all $q \in L^3$ and select the values of q for which $p(q) = 0$.

Answer: $x = 1$, $x = -3$, $x = -2/3$. (We can check this answer by expanding the product $(x-1)(x+3)(x+2/3)$ and comparing the result with the given polynomial p.)

Exercise 5.8. Make a CAS function ***rational_roots*(A)** that takes a list A of the coefficients of an integer polynomial p and returns the list of rational roots of p or the empty list if p does not have any.

Remark 5.9. For so called **monic** integer polynomials, that is, polynomials with leading coefficient one, the only possible rational roots are integers, the divisors of the constant term.

There is a classical result called **Vieta's formulas**, which relates complex or real coefficients of any polynomial to sums and products of its roots. For example, Vieta's formulas for a monic quadratic polynomial $x^2 + p\,x + q$ with roots x_1 and x_2 are $x_1 + x_2 = -p$, $x_1 \cdot x_2 = q$. For a general integer polynomial p of degree n, one of Vieta's formulas reads

$$\frac{a_0}{a_n} = (-1)^n \prod_{j=1}^{n} x_j,$$

where $x_j, j = 1, \ldots, n$, are the roots of p. This formula implies the Rational Root Theorem. (**Show this!**)

5.2 Geometry of cubic equations: Counting the number of real roots

In this subsection, based on the presentation in [15], we will first take a fresh look at the familiar quadratic equation. The key observation is that a quadratic equation $x^2 + p\,x + q = 0$ can be viewed as an equation of a line in the (p, q)-plane. Consider the set of all quadratic equations as a one-parameter set $\mathcal{F}$ of lines $l_x \equiv x^2 + p\,x + q = 0$. For example, the equation of the line l_2 is $4 + 2p + q = 0$. Thus, $\mathcal{F}$ is a set of lines in the (p, q)-plane with parameter x, $\mathcal{F} = \{l_x \mid x \in \mathbb{R}\}$. This means that we have effectively changed the role of x from being the unknown of a quadratic equation to a parameter for the corresponding line in the family $\mathcal{F}$. Note that, without loss of generality, we can use monic quadratic polynomials since dividing any quadratic equation by the leading coefficient does not affect the roots of the equation.

Next we show that this family of lines and a special curve that the family defines can be used as tools for a geometric analysis of roots of monic quadratic equations. Such heavy mathematical machinery is not needed for quadratic equations, but the

---

[3]which can be done using Horner's method or the synthetic divison algorithm

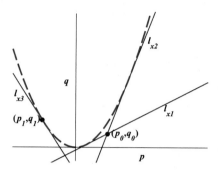

Figure 5.1. Geometry of monic quadratic equations represented by the one-parameter family of lines l_x and the envelope of the family.

same geometric idea applies to the less trivial case of cubic and, with appropriate modifications, even quartic equations.

The special curve defined by the family $\mathcal{F}$ of lines is called the **envelope**. In general, the envelope for a family of curves in the plane is a curve such that each of its points is touched tangentially by one of the curves of the family. All of these tangential points together form the entire envelope.

Finding the envelope for a family of curves. Given a family of curves $\mathcal{G} = \{g(p, q, c) = 0 \,|\, c \in \mathbb{R}\}$ in the (p, q)-plane parametrized by c, the envelope is defined by the system of equations $g(p, q, c) = 0$, $g_c'(p, q, c) = 0$. Here g_c' is the derivative of the function g with respect to c, assuming that p and q are constants.[4]

For our family of lines $\mathcal{F}$ the envelope is defined in the (p, q)-plane by the equations $x^2 + px + q = 0$, $2x + p = 0$. Using the second equation to exclude the parameter x from the first one and simplifying, we obtain the explicit equation of the envelope $q = p^2/4$. (**Show this!**) Thus, the envelope of the family $\mathcal{F}$ is the parabola $p^2 - 4q = 0$. This should ring a bell – we know that this condition identifies quadratic equations with double roots. Thus each point on the envelope corresponds to a quadratic equation with double roots.

Fig. 5.1 shows the envelope and three tangent lines to the envelope from the family $\mathcal{F}$ with parameter values $x = x_1, x = x_2$ and $x = x_3$. Tangents l_{x1}, l_{x2} pass through the point (p_0, q_0). This tells us that the equation $x^2 + p_0 x + q_0 = 0$ has two real zeros x_1, x_2. The equation $x^2 + p_1 x + q_1 = 0$ has the double root x_3 since the point (p_1, q_1) lies on the envelope.

In summary, a quadratic equation $x^2 + px + q = 0$ has two different real roots if there are two tangent lines to the envelope that pass through the point (p, q). Only one tangent passes through each point on the envelope; corresponding quadratic equations have one double root equal to the parameter of the tangent line.

Exercise 5.10.

(a) Describe the set of points in the (p, q)-plane that does not lie on any tangent to the envelope. What does this mean for the corresponding quadratic equations? Pick one of such points and check your answer.

---

[4]In multivariate calculus it is called the **partial derivative** of g with respect to c.

(b) Make a figure similar to Fig. 5.1, but with two tangents to the envelope passing through the point $(10, 20)$. Pick any point (p_1, q_1) on the envelope and include the tangent passing through this point in the figure.

We did not discover anything new about quadratic equations, but our fresh look at this familiar setting suggests an idea for constructing a similar device for geometrically analyzing the real roots of higher degree polynomials.

Consider a two-parameter family of all cubic equations of the form $x^3 + px + q = 0$. These equations can also be viewed as the family of lines $\mathcal{G} = \{l_x \mid x \in \mathbb{R}\}$ in the (p, q)-plane with parameter $x \in \mathbb{R}$. Note that, without loss of generality, we can omit the x^2 term from the general monic cubic polynomial since a change of the variable x to $X - c$ with an appropriate value of c eliminates the quadratic term from this equation.

💡 **Check Your Understanding.** Explain how the change of variables $x = X - c$ affects the roots of the original cubic polynomial. In other words, explain how the roots of the given cubic polynomial and the roots of the transformed polynomial in the variable X are related.

Exercise 5.11. Make a CAS function **remove_xsquared**(A) that takes a sequence of coefficients $A = (a_2, a_1, a_0)$ of a monic cubic polynomial and returns:

(a) The number c such that the substitution $x = X - c$ eliminates the quadratic term in the polynomial.

(b) The equation (or coefficients) of the polynomial in terms of the new variable $X = x + c$.

(c) Check the function on the polynomial $p(x) = x^3 + 3.5x^2 - 2.1x + 4.0$ and verify the relation between the roots of this polynomial and the roots of the transformed polynomial in the variable X. Use an appropriate root-finding CAS command for this part of the exercise as a "black box".

Next we will construct the main tool for our geometric exploration of the number of real roots of monic cubic polynomials, the envelope of the family of lines $\mathcal{G}$.

Exercise 5.12. Derive the equation $4p^3 + 27q^2 = 0$ for the envelope of the family $\mathcal{G}$. **Hint**: The envelope is defined by the system $x^3 + px + q = 0$ and $3x^2 + p = 0$. To exclude x from any of these equations, use the remainder of Euclidean division of the cubic polynomial by the quadratic one.

The curve $4p^3 + 27q^2 = 0$ is called a semi-cubical parabola. Fig. 5.2 shows four lines of the family $\mathcal{G}$ and the envelope. The horizontal tangent line passes through the origin. The cubic equation corresponding to the origin is $x^3 = 0$ with a triple root at $x = 0$. The point (p_0, q_0) represents the cubic equation with three real zeros equal to the parameters x_j of the tangents l_{xj}, $j = 1, \ldots, 3$, to the envelope passing through this point. Although we cannot determine the values of the roots from this geometric display, we can determine the signs of the roots: two of them are positive and one negative.

💡 **Check Your Understanding.** Explain how the signs of real roots of the equation $x^3 + p_0x + q_0 = 0$ can be determined from Fig. 5.2.

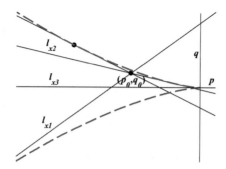

Figure 5.2. Geometry of cubic equations represented by the two-parameter family of lines and the envelope of the family.

In summary, the number of real roots of a cubic polynomial with parameters p, q is determined by the number of tangent lines to the envelope passing through the point (p, q). For any point (p, q) between the branches of the envelope, the corresponding cubic equation has three real solutions. It can be shown that any point on the envelope (except for the origin) represents a cubic equation with three real zeros such that two of them are equal. The double root is equal to $-m$, where m is the slope of the tangent passing through this point on the envelope.

Exercise 5.13.

(a) Is there a point in the (p, q)-plane such that there is no tangent line to the envelope passing through this point? Explain.

(b) Describe the region in the (p, q)-plane that corresponds to cubic equations with one real root and two complex conjugate roots.

Exercise 5.14. Make a CAS function **tangent_to_envelope**(q_0) that takes a real number q_0 and returns

(a) A figure with a plot of the semi-cubical parabola $4p^3 + 27q^2 = 0$, the point on this curve with $q = q_0$, and the tangent line passing through this point.

(b) The cubic equation without the quadratic term defined by the point of tangency found in part (a) and the double root of this cubic equation. Do not use a CAS root-finding command for this part.

(c) Test your function for $q_0 = 2$.

Answer to the test problem: (a) The equation of the tangent line passing through the point $(-3, 2)$: $q = -p - 1$; (b) The double root of the equation $x^3 - 3x + 2 = 0$ is $x = 1$.

The geometric analysis of the number of real roots can be extended to quartic polynomials in the form $p(x) = x^4 + px^2 + qx + r$ with real coefficients. (Without loss of generality, the cubic term can be eliminated by an appropriate linear change of the

Figure 5.3. Swallowtail surface. The surface represents points (p, q, r) where the polynomial $p(x) = x^4 + p\,x^2 + q\,x + r$ has a repeated root.

variable x.) However, instead of a family of lines, we have to deal with a family of planes in three-dimensional space with coordinates (p, q, r). The envelope of this family of planes is the surface shown in Fig. 5.3. Its parametric equations are obtained through some clever algebraic manipulations of the equations $p(x) = 0$, $p'(x) = 0$ and a simple change of coordinates in the space defined by the parameters p, q, r. Certain regions on this surface correspond to quartic polynomials with real roots of different multiplicity.

🔅 *Check Your Understanding.* How many real roots can a quartic polynomial have?

📖 A classic technique called **Descartes' Rule of Signs** gives an upper bound on the number of positive and negative roots of a polynomial with real coefficients. The number of positive zeros of a polynomial $p(x)$ cannot exceed the number of times the coefficients of p change sign, and the two numbers have the same parity (i.e., can differ only by a multiple of two). The number of negative zeros of a polynomial with real coefficients cannot exceed the number of times the coefficients of the polynomial $p(-x)$ change sign, and the two numbers have the same parity. Descartes' Rule of Signs is generalized by Budan's Theorem, which provides a method for counting the number of real roots of a univariate polynomial on an interval. This method is used in fast modern algorithms for real-root isolation.

5.2.1 Roots of unity. To develop a formula for finding all solutions to the equation $w^n = 1$, which are called **roots of unity**, we consider first the more general equation

$$w^n = z, \quad z = a + ib. \tag{5.2}$$

This equation can be easily solved if we convert it into polar form.

Suppose $w = \rho \exp(i\theta)$. We have to find the unknowns ρ and θ. Rewrite z as $r(a/r + i \cdot b/r)$, where $r = \sqrt{a^2 + b^2}$. Let t be an angle such that $\cos(t) = a/r$, $\sin(t) = b/r$. Then the polar form of the right-hand side of (5.2) takes the form $r \exp(it)$ and we can rewrite the equation as

$$\left(\rho \exp(i\theta)\right)^n = r \exp(i\,t),$$

or

$$\rho^n \cdot \big(\exp(in\theta) \big) = r \exp(it).$$

Two complex numbers are equal if they have equal radii and arguments that differ by a multiple of 2π. Therefore

$$\rho = \sqrt[n]{r}, \quad \theta = \frac{t + 2k\pi}{n}, \quad k = 0, 1, \ldots, n-1. \tag{5.3}$$

🔅 **Check Your Understanding.** Will we get new roots of the given equation if we allow k to take any values in $\mathbb{Z}$? Explain.

For $z = 1$ we have $t = 0$ and the general formula (5.3) gives all n^{th} roots of unity:

$$\rho = \sqrt[n]{r}, \quad \theta = \frac{2k\pi}{n}, \quad k = 0, 1, \ldots, n-1. \tag{5.4}$$

Geometrically, these numbers are vertices of a regular n-gon inscribed into a circle of radius ρ centered at the origin.

Exercise 5.15. (a) Make a CAS function ***roots_of_unity*(n)** that takes a natural number n and returns a list of all the nth roots of unity.
(b) Use the function for $n = 7$ and plot the corresponding regular heptagon inscribed into the unit circle.

📖 Complex numbers behave like coordinates on a 2D plane. Suitably paired, complex numbers form so called *quaternions*, discovered in 1843 by the Irish mathematician, astronomer, and physicist William Rowan Hamilton. John Graves, a lawyer friend of Hamilton's, subsequently showed that pairs of quaternions make *octonions*, numbers that define coordinates in an abstract 8D space.

5.3 Lab 6: Solving cubic equations using Vieta's substitution

Without loss of generality we consider cubic polynomials of the form

$$P(z) = z^3 + px + q. \tag{5.5}$$

Such polynomials are called **depressed cubic**. Solving any cubic equation can be reduced to finding the roots of a depressed cubic polynomial, as shown in Exercise 5.11.

The lab in this section is based on a clever substitution invented by F. Viète: $z = w - p/(3w)$. This substitution effectively reduces the task of finding the roots of a depressed cubic down to solving a quadratic equation for w^3. Then corresponding w-values can be found using formula (5.3) and the roots of the cubic equation are calculated from Vieta's substitution using these w-values. These algebraic steps constitute the theoretical part of the lab below and should be done manually in symbolic form. In the computational part, the derived formulas will be used to make a function that finds the roots of a depressed cubic. Our presentation of this topic is inspired by a fragment from [2].

Problem formulation. Consider a cubic equation $P(z) = 0$, where P is a depressed polynomial of the form (5.5). Do the following.

Part 1. (Theoretical) Use Vieta's substitution $z = w - p/(3w)$ to derive a quadratic equation for the new unknown $u = w^3$. Solve the quadratic equation symbolically.

Find the corresponding expressions for the w's and use them in Vieta's substitution formula to obtain the roots of P.

Part 2. (**Computational**) Construct a depressed cubic with the specified roots of your choice and use CAS to implement the algorithm in the theoretical part of the lab for this polynomial interactively. Then make a CAS function **solve_cubic(p, q)** that takes parameters p, q of a depressed cubic polynomial and returns its three roots. Test your function on the constructed depressed cubic polynomial.

Suggested directions.

Part 1. It is convenient to do this part of the lab in three steps.

Step 1. Deriving and solving a quadratic equation for the new variable u require only basic algebraic skills. Substitute $z = w - p/(3w)$ into equation (5.5) and implement algebraic operations needed to transform the left-hand side of the equation into a polynomial. The polynomial has degree six but involves only w^6 and w^3 which makes it quadratic with respect to the new variable $u = w^3$. Find u_1 and u_2.

Step 2. Convert u_1, u_2 into trigonometric form $u = \rho(\cos(t) + i \cdot \sin(t))$ and use the formula (5.3) for finding the values of $w = \sqrt[3]{u}$.

Step 3. Use the w-values found in the previous part to obtain expressions for the real and imaginary parts of the roots of the original depressed cubic equation from Vieta's substitution $z = w - p/(3w)$. This boils down to finding the real and imaginary parts of the complex number $1/w$ since the real and imaginary parts of w are already found in the previous step.

Part 2. To provide some guidance for the computational part of this lab, we show below the outputs for solving the cubic equation $z^3 - 7z + 6 = 0$ according to the algorithm described in the theoretical part. The solutions of this equation are $z = 1$, $z = 2$, and $z = -3$. The polynomial in the left-hand side of the equation is found by expanding the product $(z - 1)(z - 2)(z + 3)$.

- Equation for $u = w^3$: $u^2 + 6u + 343/27 = 0$.
- Solution: $u = -3 \pm 10i\sqrt{3}/9$.
- Trigonometric form of u : $\rho(\cos(t) \pm i \sin(t))$ with $\rho = \sqrt{381/3}$, $t \approx 2.5712158$.
- $w_k = r(\cos(\frac{t+2k\pi}{3}) \pm i \sin(\frac{t+2k\pi}{3}))$, $k = 0, 1, 2$, with $r = \sqrt[3]{\rho}$.
- Roots $z_k = w_k - p/(3w_k)$ of the given polynomial:

$$z_1 = 1.999999999 + i \cdot 10^{-9}$$

$$z_2 = -3.000000001 + i \cdot 10^{-10}$$

$$z_3 = 1.000000001 + i \cdot 10^{-9}$$

Remark 5.16. The polynomial in our demo example has integer roots. However, we used numerical approximations when calculating the cube roots of complex numbers. As a result, the imaginary parts of the solution to the cubic equation are nonzero, albeit very small.

⌑ *Check Your Understanding.* Generally, a quadratic equation for u has two distinct roots. Taking the cube root of each of them results in six complex values of w. Why do we use just three of these values? **Suggestion:** Experiment with our demo example to answer this question.

5.4 Nonnegative univariate polynomials

5.4.1 Sum of squares decompositions of nonnegative polynomials.

Definition 5.17 (Globally nonnegative univariate polynomial). A polynomial p is called **globally nonnegative** if $p(x) \geq 0$ for all $x \in \mathbb{R}$.

Theorem 5.18 (Criterion for global nonnegativity [20]). *A univariate polynomial with real coefficients is globally nonnegative if and only if it can be represented as a sum of squares of polynomials with real coefficients.*

To develop some intuition before proving the theorem, let us show that the theorem is valid for a couple of simple positive polynomials.

Example 5.19. Represent each polynomial as a sum of squares of two polynomials with real coefficients.

(a) $p(x) = x^2 + 1$

(b) $q(x) = x^2 + x + 1$

Solution.

(a) $p(x) = p_1^2(x) + p_2^2(x)$ with $p_1(x) \equiv x$, a polynomial of degree one, and $p_2(x) \equiv 1$, a polynomial of degree zero (constant).

(b) The polynomial $q(x)$ has two complex conjugate roots, $x = -1/2 \pm i\sqrt{3}/2$. Therefore it can be written as the product of two complex conjugate factors

$$q(x) = \left((x + 1/2) + i\sqrt{3}/2\right) \cdot \left((x + 1/2) - i\sqrt{3}/2\right).$$

Since $(a + ib) \cdot (a - ib) = a^2 + b^2$, we have $q = q_1^2 + q_2^2$, where

$$q_1 \equiv x + 1/2, \quad q_2 \equiv \sqrt{3}/2.$$

Proof. Clearly, a sum of squares of polynomials with real coefficients is a globally nonnegative polynomial. For the other direction, note that $p(x)$ must be of even degree with nonnegative leading coefficient. (**Why?**) To simplify notation, assume without loss of generality that the polynomial is monic. Suppose p has s complex conjugate pairs of roots z_j, z_j^*, $j = 1, 2, \ldots, s$, and k different real roots a_j, $j = 1, 2, \ldots, k$, with multiplicities m_j. Then the polynomial can be written as

$$p(x) = \prod_{j=1}^{s} (x - z_j)(x - z_j^*) \cdot \prod_{j=1}^{k} (x - a_j)^{m_j}. \tag{5.6}$$

It can be shown that all of the m_j must be even. Then each factor in the second product in (5.6) can be rewritten as $(x - a_j)^{m_j/2}(x - a_j)^{m_j/2}$ and the polynomial as

$$p(x) = \prod_{j=1}^{l} (x - w_j) \cdot \prod_{j=1}^{l} (x - w_j^*),$$

where $2l$ is the polynomial degree. Here, there are exactly s genuine complex w's in each product and equal values of w's may occur. Since the two products are complex conjugates of each other, separating the real and the imaginary parts for each product (like in Example 3 (b)), we obtain

$$p(x) = (q(x) + ir(x))(q(x) - ir(x)).$$

Here $q(x) \equiv Re(\prod_{j=1}^{l}(x - w_j))$ and $r(x) \equiv Im(\prod_{j=1}^{l}(x - w_j))$. Thus $p(x) = q^2(x) + r^2(x)$.   □

The next example illustrates an application of the theorem.

Example 5.20. Write the polynomial $p(x) = (x^2 + x - 2)^2 \cdot (x^2 + x + 1)$ as the sum of squares of two polynomials.

Solution. Clearly the polynomial is nonnegative. It has two real double roots and two complex conjugate roots. Since the real roots are already separated in the first factor of the given polynomial, it can be rewritten as

$$p(x) = \left((x^2 + x - 2)\left(x + 1/2 + i\frac{\sqrt{3}}{2}\right)\right) \cdot \left((x^2 + x - 2)\left(x + 1/2 - i\frac{\sqrt{3}}{2}\right)\right).$$

This is the product of complex conjugate expressions with real part $a = (x^2 + x - 2)(x + 1/2)$ and imaginary part $b = (x^2 + x - 2)\sqrt{3}/2$. Thus

$$p(x) = \left((x^2 + x - 2)\left(x + \frac{1}{2}\right)\right)^2 + \left(\frac{\sqrt{3}}{2}(x^2 + x - 2)\right)^2.$$

To check the solution, use CAS to expand the given polynomial and its decomposition and compare the two expansions.

For multivariate polynomials the statement of the above theorem is not generally true, that is, there are nonnegative polynomials in m variables with real coefficients (termed positive semidefinite) that cannot be decomposed into a sum of squares of polynomials with real coefficients. However, in 1927, Emil Artin proved that a real nonnegative polynomial in m variables is a sum of squares of rational functions, thus solving Hilbert's 17th problem. The existence (nonexistence) of sum of squares (SOS) decompositions of real multivariate polynomials is still an active area of current mathematical research. In optimization theory, SOS decompositions are used for proving nonnegativity of polynomials.

5.4.2 Mini project: Decomposing a nonnegative univariate polynomial into a sum of squares. Consider the polynomial $p(x) = x^8 - 3x^7 - 9x^6 + 22x^5 + 27x^4 - 27x^3 - 23x^2 - 24x + 36$. Find the decomposition of $p(x)$ into a sum of squares of two polynomials, if possible.

Suggestion. Use the function *rational_roots* made in Exercise 5.8. Then, using the function *synthetic_div* made in Exercise 5.6, reduce the degree of the given polynomial by the number of rational roots and follow the steps of Example 5.20. Alternatively, use an appropriate root-finding CAS function for finding the roots of the given polynomial and continue from there.

5.5 Glossary

- **Roots of a univariate polynomial** – The values of a variable for which the given polynomial is equal to zero.

- **The Fundamental Theorem of Algebra** – Every non-constant univariate polynomial with complex coefficients has at least one complex root.

- **Multiplicity of a root of a polynomial** – One more than the number of its derivatives that have the same root.

- **Vieta's formulas** – Relate coefficients of a polynomial and its roots.

- **Integer polynomial** – A polynomial with integer coefficients.

- **Roots of unity** – Solutions $w \in \mathbb{C}$ to the equation $w^n = 1$.

- **Depressed cubic polynomial** – A polynomial in the form $x^3 + px + q$. Any cubic polynomial can be reduced to this form by an appropriate linear change of variables.

- **Globally nonnegative polynomial** – A polynomial $p(x)$ with real coefficients such that $p(x) \geq 0$ for all $x \in \mathbb{R}$.

6

Topics in Algebra: Bivariate Systems of Polynomial Equations

In this chapter we consider only systems of two polynomial equations of degree one or two in two variables. We will present both geometric and algebraic approaches to find solutions of such systems. From a geometric standpoint, if a nonlinear polynomial system of two equations in two unknowns defines two curves on the plane, then finding the real solutions of such a system means finding the points of intersection of these curves. To solve a system of two polynomial equations algebraically one needs some elimination theory. In the case of a system of linear equations, it is a classic Gaussian elimination technique. For polynomials of higher degree, the elimination can be done using the relatively new and advanced Gröbner bases techniques. These techniques are beyond the scope of this chapter.

There is a different classical approach to solving systems of two polynomial equations that employs the so called **resultant** of two polynomials. The resultant of two polynomials is the determinant of a special matrix called the Sylvester matrix, which is constructed from the coefficients of the polynomials. The method of resultants works for finding both real and complex roots. The most important property that makes the resultant useful in analyzing and solving systems of polynomial equations is that **the resultant of two univariate polynomials equals zero if and only if the polynomials have a common root**. Resultants provide the oldest proof that algorithms for solving polynomial systems exist. In this chapter we introduce this powerful mathematical tool constructively and demonstrate by simple examples how it can be used. The general approach and rigorous explanation of why resultants work are beyond the scope of this chapter.

In addition to solving systems of two polynomial equations, the resultant can be used in a process called "implicitization," which is a method to convert parametric equations of a curve into an equivalent implicit form defined by a polynomial equation. It is known that such a conversion is always possible for curves defined by rational functions of the form $x = p(t)/r(t)$, $y = q(t)/r(t)$, where p, q, r are polynomials. Such curves are called **rational**. It is also known that the inverse conversion, called a **rational parametetrization**, is not always possible. That is, there are plane curves defined implicitly by polynomial equations that do not have a rational parametrization. This chapter includes a brief description of the resultant-based implicitization method and a lab where this technique will be used. Implicitization and its advanced versions for parametric surfaces have applications in computer vision and Computer Aided Graphics Design (CAGD).[1]

🖘 The ubiquitous need for solving polynomial equations is at the heart of **algebraic geometry**, a branch of mathematics that studies the zeros of multivariate polynomials. This classical field was revitalized in the 1960s when Buchberger and Hironaka discovered new algorithms for manipulating systems of polynomial equations using the concept of a Gröbner basis. The development of computers fast enough to execute these algorithms made it feasible to investigate complicated polynomial systems that would have been impossible to do by hand. These computations greatly enhanced applications of algebraic geometry in mathematics and various sciences. Alexander Grothendieck, who received the Fields medal for advances in algebraic geometry and two other fields of mathematics, is considered to have been the leading figure in the development of modern algebraic geometry.

6.1 Linear systems of two equations

In this section we introduce the main tools for this chapter – the Sylvester matrix and resultant – for a system of two linear equations with two unknowns. This will help to develop some intuition for these nontrivial mathematical constructions that we will use later in this chapter for solving nonlinear systems.

Consider a system of two linear equations with one unknown

$$a_1 x + d_1 = 0$$
$$a_2 x + d_2 = 0. \tag{6.1}$$

Substituting $x = -d_2/a_2$ found from the second equation into the first one, we obtain $a_2 d_1 - a_1 d_2 = 0$. This solvability condition for the system (6.1) can be equivalently written as

$$\det \begin{bmatrix} a_1 & d_1 \\ a_2 & d_2 \end{bmatrix} = 0. \tag{6.2}$$

The matrix on the left-hand side of this equation is the Sylvester matrix of the polynomials $p_j(x) = a_j x + d_j$, $j = 1, 2$, that define the system (6.1). Its determinant is the **resultant**.

Next consider a system of two linear equations with two unknowns

$$p_1(x, y) = a_1 x + b_1 y - c_1 = 0$$
$$p_2(x, y) = a_2 x + b_2 y - c_2 = 0. \tag{6.3}$$

---

[1]See, for example, [**17**].

Let us assume for a moment that the only unknown we are interested in is x and consider y as a parameter. Then we can rewrite p_j as the univariate polynomials $a_j x + d_j$, where we use the notation $d_j = b_j y - c_j$. Now the system becomes identical to (6.1). The criterion (6.2) for the existence of x-values satisfying the given system of linear equations reads

$$\det \begin{bmatrix} a_1 & b_1 y - c_1 \\ a_2 & b_2 y - c_2 \end{bmatrix} = 0.$$

The matrix on the left-hand side of this equation is the Sylvester matrix, and its determinant is the resultant of the polynomials p_j **with respect to** x. Typically, the notation $\operatorname{Res}(p, q, x)$ is used for the resultant of two polynomials p, q with respect to x. After a simple algebraic manipulation, the equation yields $\Delta \cdot y = \Delta_y$ with

$$\Delta = \det \begin{bmatrix} a_1 & b_1 \\ a_2 & b_2 \end{bmatrix}, \quad \Delta_y = \det \begin{bmatrix} a_1 & c_1 \\ a_2 & c_2 \end{bmatrix}. \tag{6.4}$$

Similar arguments lead to the solvability condition $\operatorname{Res}(p_1, p_2, y) = 0$ when x is treated as a parameter.

💡 *Check Your Understanding.*

(a) Derive equations (6.4).

(b) Derive $\operatorname{Res}(p_1, p_2, y)$.

The solvability condition $\operatorname{Res}(p_1, p_2, y) = 0$ can be written as $\Delta \cdot x = \Delta_x$ with $\Delta_x = \det \begin{bmatrix} c_1 & b_1 \\ c_2 & b_2 \end{bmatrix}$. If $\Delta \neq 0$, the two solvability conditions give the unique solution of the system:

$$x = \frac{\Delta_x}{\Delta}, \quad y = \frac{\Delta_y}{\Delta}. \tag{6.5}$$

This is known as **Cramer's Rule** for solving a linear system of two equations with two unknowns, provided $\Delta \neq 0$.

📖 The formula (6.5) generalizes to a system of linear equations with as many equations as unknowns and a nonzero determinant of a matrix formed from the coefficients of the unknowns. Cramer's Rule is not practical for large systems.

Remark 6.1 (Geometric meaning of the system $p_j(x, y) = 0$, $j = 1, 2$). Since each of the equations of the system defines a line on the plane, solving the system means finding the point of intersection of the two lines, if possible.

💡 *Check Your Understanding.*

(a) What is the geometric meaning of the conditions $\Delta = 0$, $\Delta_x \neq 0$ for the two lines defined by the system (6.3)? (A linear system that satisfies these conditions is called **inconsistent**.)

(b) What is the geometric meaning of the conditions $\Delta = 0$, $\Delta_x = 0$ for the two lines defined by the system (6.3)? (A linear system that satisfies these conditions is called **indeterminate**.)

Exercise 6.2.

(a) Solve the system

$$2x - 3y = 1,$$
$$x + y = 2.$$

using Cramer's Rule.

(b) Make a CAS function ***my_cramer*(coef1, coef2)** that takes the lists of coefficients $coef1, coef2$ of two linear polynomials in two variables $p_j(x, y)$, $j = 1, 2$, and returns the unique solution of the system of linear equations $p_j(x, y) = 0$ using Cramer's Rule or prints one of the following two statements: "The system is inconsistent" or "The system is indeterminate". **Suggestion**: Use an appropriate *if-elif-else* conditional statement to control the flow of the function execution for each of the three possible scenarios.

6.2 Nonlinear systems of polynomial equations: Motivating example

Before discussing the methods for solving systems of nonlinear equations in the next section, let us consider a simple motivating problem from robotics.

6.2.1 Motivating example: Inverse kinematics of a two-link planar manipulator. In applications, algebraic methods for solving an applied problem typically include three steps:

(1) Algebraization – translating the problem into the language of algebra

(2) Solving the algebraic problem

(3) Interpreting the solution in the context of the applied problem

We will illustrate these three steps in the following kinematic analysis[2] of a very simple two-link robot arm. The solution method relies on the specific form of the algebraic model for the problem and does not involve any general standard techniques.

Consider the schematic picture of a two-link planar manipulator in Fig. 6.1a and an abstract geometric model of this device in Fig. 6.1b. Our mathematical analysis of this model will allow us to determine two parameters, $q_j \in [0, 2\pi)$, $j = 1, 2$, which ensure that the design specifications are met. Any pair of these parameters define what is called in robotics the **configuration** of this two-link robot arm.

Formulating the inverse kinematic problem for the planar two-link robot. Consider the model depicted in Fig. 6.1b. Given the lengths of the two links and the coordinates of the target point P, find the angles for the links that position the tip of the second link (the "hand") at the target point P.

Note that if the target point lies within the range of the manipulator's end, and the two links do not have to be fully extended to reach this point, then there are two different solutions corresponding to "elbow-up" (dashed lines) and "elbow-down" (solid lines) in Fig. 6.1b.

---

[2]Kinematic analysis is the study of motion of a machine or mechanism without considering the forces that cause the motion. It allows designers to determine parameters like the position, displacement, rotation, speed, velocity, and acceleration of a given mechanical structure.

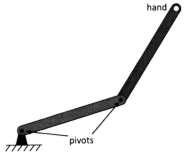

(a) Schematic of a two-link robot manipulator.

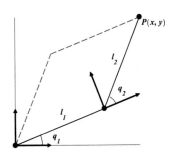

(b) A geometric abstraction of a two-link robot manipulator.

Figure 6.1. Two-link planar manipulator.

Notation.

- v_1 – vector of length l_1 along the first link with its tail at the first joint (pivot)

- v_2 – vector of length l_2 along the second link with its tail at the second joint

- $r = \sqrt{x^2 + y^2}$ – distance from the origin to the target point

We assume that the angles of rotation of the links are not restricted by the robot design, that is, $q_1, q_2 \in [0, 2\pi)$.

Geometric model. The goal is to find the angle values q_j such that $v_1 + v_2 = \overrightarrow{OP}$. Using basic trigonometry, we can write the coordinates of the vectors v_1, v_2 in terms of the parameters of the problem and the unknown angle values:

$$v_1 = \begin{bmatrix} l_1 \cos(q_1) \\ l_1 \sin(q_1) \end{bmatrix}, \qquad v_2 = \begin{bmatrix} l_2 \cos(q_1 + q_2) \\ l_2 \sin(q_1 + q_2) \end{bmatrix}.$$

Now the vector equation $v_1 + v_2 = \overrightarrow{OP}$ can be restated in coordinate form as

$$l_1 \cos(q_1) + l_2 \cos(q_1 + q_2) = x,$$
$$l_1 \sin(q_1) + l_2 \sin(q_1 + q_2) = y.$$

Using the notation $c_j \equiv \cos(q_j)$, $s_j \equiv \sin(q_j)$, along with the sum and difference identities for sine and cosine, we rewrite this system as the two algebraic equations

$$\begin{cases} l_1 c_1 + l_2(c_1 c_2 - s_1 s_2) = x, \\ l_1 s_1 + l_2(s_1 c_2 + s_2 c_1) = y. \end{cases} \tag{6.6}$$

These two equations, combined with the Pythagorean trigonometric identities

$$c_j^2 + s_j^2 = 1, \ j = 1, 2, \tag{6.7}$$

represent a kinematic model of the two-link planar robot. The model is a system of four nonlinear algebraic equations for the unknowns c_j, s_j.

Sketch of solution. Notice that if $l_1 + l_2 < r$, the system is inconsistent, that is, there is no solution. If $l_1 + l_2 = r$ then the solution is trivial. (**Find it!**) Let us assume that $l_1 + l_2 > r$, as in Fig. 6.1b. Squaring each of the equations in (6.6) and then adding the results and simplifying yields an equation that can be solved explicitly for c_2. In addition, (6.7) can be solved for $s_2 = \pm\sqrt{1 - c_2^2}$. Substituting these formulas for c_2 and s_2 into (6.6) gives us a linear system for s_1 and c_1.

☼ *Check Your Understanding.* Derive the formula for c_2.

Exercise 6.3. Assuming that c_2, s_2 are known, rewrite the system (6.6) as a linear system with respect to the variables c_1, s_1 and solve it symbolically using Cramer's Rule. **Suggestion**: Use the CAS function *my_cramer* made in Exercise 6.2 or solve the system manually.

Model using Euler's formula. The vectors $\mathbf{v}_1$, $\mathbf{v}_2$ can be written as $\mathbf{v}_1 = l_1 \exp(iq_1)$, $\mathbf{v}_2 = l_2 \exp\big(i(q_1 + q_2)\big)$ and the target point as $P = r \exp(iq_0)$, where $q_0 = \arg(x + iy)$. Now the model of the problem takes the form

$$l_1 \exp(iq_1) + l_2 \exp\big(i(q_1 + q_2)\big) = r \exp(iq_0),$$

or

$$\exp(iq_1)\big(l_1 + l_2 \exp(iq_2)\big) = r \exp(iq_0). \tag{6.8}$$

The mathematical beauty of the complex form of the model is in using only one equation instead of four real equations. Notice that two Pythagorean identities hold automatically due to Euler's formula.

The solution can be found by equating the moduli and arguments of the left- and right-hand sides of equation (6.8). Equating squares of the moduli, we obtain

$$(l_1 + l_2 c_2)^2 + l_2^2 s_2^2 = r^2.$$

☼ *Check Your Understanding.* Derive the formula for c_2 from this equation.

Then, as before, using the Pythagorean identity, one can calculate two corresponding values of $\sin(q_2)$ provided $\cos(q_2) \neq \pm 1$. Two values of the sine function and one value of the cosine uniquely identify two values $q_2^{(1)}$, $q_2^{(2)}$ of the angle q_2. Equating the arguments of the left- and right-hand sides in (6.8), we obtain an elementary equation for finding q_1[3]:

$$q_1 + \arg\big((l_1 + l_2 c_2) + il_2 s_2\big) = q_0.$$

Notice that if $\cos(q_2) \neq \pm 1$, then this equation has two solutions for q_1 corresponding to two different values of s_2.

6.2.2 Mini project: The inverse kinematic problem for a two-link robot arm.

---

[3]Recall that the argument of the product of two complex numbers is the sum of the arguments of the factors.

Problem formulation. Write two CAS functions:

(1) **robot_configuration**(l_1, l_2, pt) that takes

    (a) the lengths l_j, $j = 1, 2$, of the two links of the planar robot manipulator depicted in Fig. 6.1b,

    (b) the coordinates of the target point $P(x, y)$

and returns the angle value(s) q_j that position the hand of the manipulator at the target point or the number -1 if there is no solution.

(2) **robot_plot**(l_1, l_2, pt) that takes the same parameters, calls the function **robot_configuration**, and returns the string "No solution." if the output is -1, or a figure similar to Fig. 6.1b defined by the solution of the problem.

Suggested directions.

Step 1. To make the function **robot_configuration**, use either of the two mathematical models presented in this section. Use an *if-elif-else* conditional statement to encode three versions of the output:

- the number -1, if $l_1 + l_2 < r$
- the unique solution, if $l_1 + l_2 = r$
- two solutions, $q_1^{(j)}$, $q_2^{(j)}$, $j = 1, 2$, if $l_1 + l_2 > r$

Step 2. Test the function in Step 1 on the data $l_1 = 5, l_2 = 17, pt = (18.0, 12.0)$. Interpret your answer in the context of this applied problem.

Formal answer: $q_1^{(1)} = 0.24871$, $q_2^{(1)} = 0.43734)$, $q_1^{(2)} = 0.92730$, $q_2^{(2)} = -.43734$.

Step 3. Plot the abstract geometric configurations of the manipulator and the target point defined by the data in Part 2 in one figure. Your figure will be similar to Fig. 6.1b.

6.3 Solving nonlinear polynomial systems

In this section we show two approaches to finding solutions for systems of two polynomial equations of degree one or two in two unknowns. Geometrically, such a system can define two lines, two conics, or a conic and a line. (Recall that the three non-degenerate types of conic sections are hyperbola, parabola, and ellipse.) In turn, finding real solutions of such systems translates to finding the intersection points of the curves defined by the given equations. Some of the systems involve parameters, and we will analyze how the number of real solutions depends on the parameter values.

6.3.1 Geometric approach to solving bivariate polynomial systems.

The geometric approach is useful for developing some intuition in finding real solutions to bivariate polynomial systems that involve parameters. However, it requires an individual treatment of every particular system, similar to the tricks we used to solve the applied problem from robotics.

    Some problems in this section involve so called **reducible polynomials**. These are polynomials with real coefficients that can be factored into a product of non-constant polynomials with real coefficients. Factoring a reducible polynomial significantly simplifies the solution procedure, as we will demonstrate in the next example.

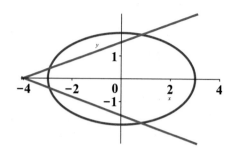

Figure 6.2. Graphs of two polynomials in Example 6.4: case of four solutions.

Example 6.4. Analyze the number of solutions to the system $p(x, y) = 0$, $q(x, y) = 0$ with

$$p(x, y) = \frac{x^2}{9} + \frac{y^2}{4} - 1, \quad q(x, y) = a(x + 4)^2 - y^2.$$

Then find solutions for all values of the parameter $a \in \mathbb{R}$ that ensure solvability of the system.

Solution. The first equation defines an ellipse. The second equation defines the empty set for $a < 0$. (**Why?**) For $a > 0$, the polynomial q is reducible; it defines the pair of lines $y = \pm c(x + 4)$, where $c = \sqrt{a}$. Fig. 6.2 shows the graphs of both equations of the system with $a = 3/8$. In this case, the system has four solutions. The picture suggests that there can be two, four, or zero solutions as the parameter c varies in the interval $[0, \infty)$.

The table below specifies the number of solutions N for different values of parameter a. The notation a^* represents the value of the parameter when the corresponding two lines are tangent to the ellipse.

	$a = 0$ or $a = a^*$	$0 < a < a^*$	$a > a^*$
N	2	4	0

Due to the problem symmetry, it suffices to solve the reduced system $p(x, y) = 0$, $y = c(x + 4)$. We can substitute $y = c(x + 4)$ into the polynomial p and solve the resulting quadratic equation for x manually or using CAS to obtain

$$x = 6 \cdot \frac{-6c^2 \pm \sqrt{4 - 7c^2}}{9c^2 + 4}.$$

Analyzing these expressions yields $c^* = \sqrt{a^*} = 2\sqrt{7}/7$. (**Explain this.**)

Exercise 6.5.

(a) What is the geometric meaning of the number c^* in Example 6.4?

(b) Derive the quadratic equation for solutions of the system in Example 6.4 and use it to find solutions of the given system for $c = c^*$, $c = 0$, and any c-value of your choice in the interval $(0, c^*)$.

(c) Plot the two given polynomials and your solution points for $c = c^*$.

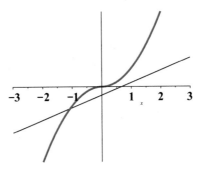

Figure 6.3. Graphs of the left- and right-hand sides of the modified
equation from Exercise 6.6 for a certain value of the parameter a.

Exercise 6.6. Geometrically analyze the number of solutions of the equation $9x|x| +$
$(a-5)x+4 = 0$ depending on the values of the parameter a. Determine the parameter
value for which the equation has exactly two solutions and find these solutions.
Suggestion: To develop some intuition, rewrite the equation as $9x|x| = (5-a)x-4$ and
plot the curve defined by the left-hand side of this equation along with lines defined by
the right-hand side for different values of the parameter a in one figure. Fig 6.3 shows
one such figure.

Remark 6.7. Strictly speaking, the expression $9x|x|$ is not a polynomial, but it becomes
polynomial if you restrict x to positive or negative values. Thus, it is a piecewise polynomial.

Exercise 6.8.

(a) Analyze the number of solutions of the system $p(x, y) = 0$, $q(x, y) = 0$ with
$$p(x, y) = x^2 + y^2 - 1, \ q(x, y) = x\,y - c,$$
depending on the values of the parameter c.

(b) Choose one value of the parameter c for which the system has more than one solution. Find these solutions and plot the given polynomials and solution points.

Exercise 6.9. Consider the system $p(x, y) = 0$, $q(x, y) = 0$ with
$$p(x, y) = x^2 - 2x - y, \ q(x, y) = a^2 + x^2 + y^2 - 2a\,y - 2x.$$

(a) Show that the second equation defines a circle. Plot the parabola $p(x, y) = 0$ and
the circle for a few different parameter values in one figure. Describe how the circle
moves when the parameter a changes.

(b) Solve the system for $a = -2$.

(c) What is the largest possible number of real solutions for this system?

6.3.2 Algebraic approach to solving bivariate polynomial systems. In
this section we introduce a simple version of an algebraic method for solving bivariate polynomial systems of two equations of degree one or two with real coefficients.

The method employs the **resultant** of two polynomials that is the determinant of a so called Sylvester matrix. Two Sylvester matrices for the general linear system of polynomials in two variables have been constructed in Section 6.1. We used corresponding resultants to derive Cramer's Rule for finding the solution of any such system of linear equations with nonzero determinant. In the next subsection we will show the details of constructing Sylvester matrices for systems that involve polynomials of degree two. For bivariate system two different Sylvester matrices can be constructed, and each corresponding resultant is a polynomial in only one variable. Thus, the resultant is a valuable mathematical tool that can be used to eliminate one of the variables in polynomial systems.

Constructing Sylvester matrices. In general, a **Sylvester matrix** of univariate polynomials p and q is a quadratic $N \times N$ matrix, where

$$N = \text{degree of } p + \text{degree of } q.$$

We will show the construction of Sylvester matrices for symbolic univariate and bivariate systems. Since we consider here only systems of polynomials of order one or two, the size of needed Sylvester matrices does not exceed four. The reader will need CAS assistance for finding the determinants of Sylvester matrices.

The next example shows two basic cases of constructing Sylvester matrices needed in this section.

Example 6.10. Construct Sylvester matrices for the following pairs of univariate polynomials.

(a) $p = a_1 x^2 + b_1 x + c_1$ and $q = b_2 x + c_2$

(b) $p_j = a_j x^2 + b_j x + c_j$, $j = 1, 2$.

Solution.

(a)

$$S = \begin{bmatrix} a_1 & b_1 & c_1 \\ b_2 & c_2 & 0 \\ 0 & b_2 & c_2 \end{bmatrix}$$

(b)

$$S = \begin{bmatrix} a_1 & b_1 & c_1 & 0 \\ 0 & a_1 & b_1 & c_1 \\ a_2 & b_2 & c_2 & 0 \\ 0 & a_2 & b_2 & c_2 \end{bmatrix}$$

In the next example we consider two specific bivariate polynomials and construct their corresponding Sylvester matrices. Each of the matrices depends only on one of the unknowns. The main idea of the resultant method is to (temporarily) assume that one of the variables is a parameter. Then the two given polynomials can be treated as univariate in the other variable. Coefficients of these polynomials depend on the variable chosen as the parameter.

Example 6.11. Let $p = x^2 - xy + \frac{4}{9}y^2 + x - 6$, $q = 3xy - y^2$. Assuming that x is a parameter, we rewrite the first polynomial as quadratic in y : $p = \frac{4}{9}y^2 + (-x)y + $

$(x^2 + x - 6)$. The second equation is already quadratic in y with zero free term. This missing term is represented by zero in the 4×4 Sylvester matrix, which is constructed according to the pattern in part (b) of Example 6.10:

$$S_1 = \begin{bmatrix} 4/9 & -x & x^2 + x - 6 & 0 \\ 0 & 4/9 & -x & x^2 + x - 6 \\ -1 & 3x & 0 & 0 \\ 0 & -1 & 3x & 0 \end{bmatrix}.$$

Assuming that y is a parameter, we rewrite these polynomials as $p = x^2 + (1 - y)x + (\frac{4}{9}y^2 - 6)$, $q = (3y)x - y^2$. Note that q is linear polynomial in x, and we construct the 3×3 Sylvester matrix according the pattern in part (a) of Example 6.10:

$$S_2 = \begin{bmatrix} 1 & 1 - y & \frac{4}{9}y^2 - 6 \\ 3y & -y^2 & 0 \\ 0 & 3y & -y^2 \end{bmatrix}.$$

Equating a resultant (the determinant of Sylvester matrix) to zero, one can find the values of the variable chosen as parameter satisfying the given system. Then using any of the two given equations, corresponding values of the second unknown can be found. Most general purpose CASs have a built-in function for computing the resultant and/or Sylvester matrix of two polynomial equations. Using CAS assistance, we find for S_1 in the example above

$$\text{Res}(p, q, y) = \det S_1 = 2x^4 + 3x^3 - 17x^2 - 12x + 36.$$

Solving the equation $\text{Res}(p, q, y) = 0$ with CAS help, we obtain the following x-values: $-2, 3/2, 2, -3$. Corresponding y-values can be easily found from the equation $q(x, y) = 0$. Alternatively, we can start with $\text{Res}(p, q, x) = \det S_2$, find the y-values $0, 0, 9/2, -6$ as the roots of this resultant, and then determine the corresponding x-values.

⚡ *Check Your Understanding.* Implement the resultant method starting with $\text{Res}(p, q, x) = \det S_2$.

6.3.3 More resultants by examples.

Resultant method in a nutshell. Given a system of polynomial equations $p(x, y) = 0$, $q(x, y) = 0$:

(1) Find $\text{Res}(p, q, x)$. The resultant is a polynomial in the variable y.

(2) Solve the equation $\text{Res}(p, q, x) = 0$ to obtain y-values y_j, $j = 1, \ldots, k$.

(3) For each y_j solve any of the two given equations for the corresponding value(s) of x.

Alternatively, one can start with $\text{Res}(p, q, y)$ and follow this algorithm with obvious modifications.

Solution of the system in the next example can be found just by substitution of the variable x from the second equation into the first one. However, we included this elementary example to emphasize the systematic nature of the resultant method as opposed to ad hoc tricks that work only for certain specific cases.

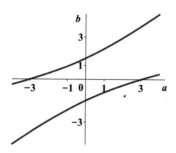

Figure 6.4. Hyperbola in the space of parameters a, b defined by the zeros of the resultant in Example 6.13.

Example 6.12. Consider the system $p_a(x) = 0$, $q(x) = 0$ with

$$p_a(x) = a\,x^2 - x + 6, \quad q(x) = x - 1.$$

Use the resultant to find the value of a that makes the system consistent.

Solution. Here is the Sylvester matrix for the two given polynomials:

$$S(a) = \begin{bmatrix} a & -1 & 6 \\ 1 & -1 & 0 \\ 0 & 1 & -1 \end{bmatrix}.$$

We found $\mathrm{Res}(p_a, q, x) = \det S = a + 5$. The system is solvable for $a = -5$.

Example 6.13. Consider the polynomial system $p_b(x) = 0$, $q_a(x) = 0$ with

$$p_b(x) = x^2 + b\,x - 1, \quad q_a(x) = x^2 + a\,x - 4.$$

(a) Find the resultant $\mathrm{Res}(p_b, q_a, x)$ of the polynomials.

(b) Plot the curve $\mathrm{Res}(p_b, q_a, x) = 0$ in ab-space. (Notice that each point on the curve defines a solvable system.)

(c) Pick a point (a_0, b_0) on the curve found in part (b), plot the corresponding two parabolas defined by the given polynomials with these parameters, and find the solution of the system with $a = a_0$, $b = b_0$.

Solution.

(a) The Sylvester matrix for the two given polynomials is

$$S(a, b) = \begin{bmatrix} 1 & b & -1 & 0 \\ 0 & 1 & b & -1 \\ 1 & a & -4 & 0 \\ 0 & 1 & a & -4 \end{bmatrix}.$$

(b) Using CAS assistance, we find $\mathrm{Res}(p_b, q_a, x) = \det(S) = -4b^2 + 5a\,b - a^2 + 9$. The equation $-4b^2 + 5a\,b - a^2 + 9 = 0$ defines a conic. Since the discriminant of this equation is $5^2 - 4 \cdot (-1) \cdot (-1) > 0$, this is a hyperbola in the ab-plane of parameters (see Fig. 6.4). Each point on the hyperbola defines a solvable system of equations in this example.

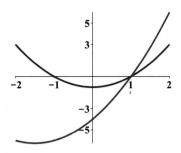

Figure 6.5. Plots of the two polynomials in Example 6.13 with parameters $a = 3$, $b = 0$.

(c) Solving the equation $\text{Res}(p_a, q_b, x) = 0$ for a, we obtain $b = 5a/8 \pm 3\sqrt{a^2 + 16}/8$. Choose, for instance, $a = 3$. Then $b = 0$ or $b = 15/4$. Fig 6.5 shows the graphs of the two parabolas with parameters $a = 3$ and $b = 0$. The solution to this system is $x = 1$.

Example 6.14. Use the method of resultants to solve the polynomial system in two unknowns $p(x, y) = 4x - y - 4 = 0$, $q(x, y) = 3y^2 - 4xy - x^2 - 3 = 0$.

Solution. Fig. 6.6 shows the plots of the two given polynomials. Note that since the first equation is linear, the system can be reduced to a single quadratic equation and easily solved manually. However, we want again to demonstrate the generality of the resultant method. We consider the variable y as a parameter and construct the Sylvester matrix $S(y)$ to obtain

$$S(y) = \begin{bmatrix} -1 & -4y & 3y^2 - 3 \\ 4 & -y - 4 & 0 \\ 0 & 4 & -y - 4 \end{bmatrix}.$$

The determinant of $S(y)$ is:

$$\text{Res}(q, p, x) = 31y^2 - 72y - 64.$$

Zeros of the resultant are $y = (36 \pm 4\sqrt{205})/31$. Now, we obtain $x = 1 + y/4$ from the equation $p(x, y) = 0$ and calculate the corresponding x-values: $x = (40 \pm \sqrt{205})/31$.
Answer: $x = (40 \pm \sqrt{205})/31$, $y = (36 \pm 4\sqrt{205})/31$, or approximately, $(1.752, 3.009), (0.828, -0.686)$.

6.3.4 Solving polynomial systems by the resultant method.
Now, after developing some feel for resultants, let us take a look at some polynomial systems in two variables and use resultants and CAS assistance to solve them.

Exercise 6.15. Use CAS assistance to interactively implement the method of resultants and find approximate (float) solutions to the system $p(x, y = 0$, $q(x, y) = 0$ with polynomials p and q given below.[4] For each system make a figure with plots of the two given implicit curves. (One of the figures will be similar to Fig. 6.7.)

(a) $q(x, y) = 2x^2 - 5xy + 4y^2 - 1$, $p(x, y) = 3y^2 - 3 + (6y - 4)x - x^2$.

---

[4]CAS can find closed-form roots of the resultants in this exercise but they look ugly.

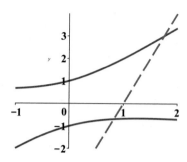

Figure 6.6. Graphs of two polynomials in Example 6.14.

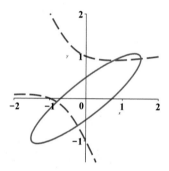

Figure 6.7. Graphs of two polynomials from one of the systems in Exercise 6.15.

(b) $q(x, y) = y + 4 + yx - 4x^2$, $p(x, y) = 3y^2 - 3 - 4xy - x^2$.

(c) $q(x, y) = x^2 + y^2 - 10$, $p(x, y) = x^2 + xy + 2y^2 - 16$.

The next problem is on solving just one equation but the equation is trigonometric, not polynomial. We will transform the problem to a polynomial system by choosing appropriate new variables, and then solve it using the resultant method. Technically, the problem is more involved than what we have done so far in this section. Its purpose is to demonstrate the power of the resultant method.

Example 6.16 (Solving a trigonometric equation by reduction to an algebraic system). Find all solutions to the equation

$$\sin^3(t) + \cos(3t) = 0 \qquad\qquad (6.9)$$

on the interval $[-\pi, \pi]$.

Solution. First, let us transform the equation to one with trigonometric functions of the same argument. Using the first formula in (2.9) we write

$$\cos(3t) = \frac{\exp(3it) + \exp(-3it)}{2} = \frac{(\exp(it))^3 + (\exp(-it))^3}{2}.$$

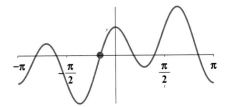

Figure 6.8. Graph of the function f in Example 6.16. The solid red circle marks one of the zeros found by reducing the trigonometric equation $f(t) = 0$ to an algebraic system.

Next we use the algebraic identity $a^3 + b^3 = (a+b)(a^2 - ab + b^2)$ with $a = \exp(it)$, $b = \exp(-it)$ to transform the right-hand side of this equation to

$$\cos(t)\big(\exp(2it) - 1 + \exp(-2it)\big) = \cos(t)(2\cos(2t) - 1).$$

Finally, the trigonometric identity $\cos(2t) = 2\cos^2(t) - 1$ yields

$$\cos(3t) = 4\cos^3(t) - 3\cos(t).$$

Using this result, the Pythagorean identity, and the notation $x = \sin(t)$ and $y = \cos(t)$, the given trigonometric equation can be recast into the polynomial system $p(x, y) = 0$, $q(x, y) = 0$, where

$$p(x, y) = x^3 + 4y^3 - 3y, \quad q(x, y) = x^2 + y^2 - 1.$$

The resultant of this system, found with CAS assistance, is

$$\mathrm{Res}(p, q, x) = 17y^6 - 27y^4 + 12y^2 - 1 = 0.$$

This is a cubic equation in y^2. Any general purpose CAS can solve the cubic equation exactly, but to avoid cumbersome expressions we choose approximations of these zeros of the resultant and obtain six values of the unknown y:

$$y \approx \pm 0.32807, \pm 0.83741, \pm 0.88281.$$

We will find x corresponding to just one of these y-values to illustrate how the solution to the problem can be completed. Consider, for instance, $y_5 \approx 0.88281$. From the equation $p(x, y) = 0$ we find $x_5 = \sqrt[3]{3y_5 - 4y_5^3} \approx -0.46973$. Since $\sin(t_5) = x_5 < 0$ and $\cos(t_5) = y_5 > 0$, the angle t_5 lies in the fourth quadrant and therefore can be found as $t_5 = \arcsin(-0.46973) \approx -0.48898$. The approximate value of the trigonometric polynomial in equation (6.9) at $t = t_5$ is 0.0000284224, which is convincingly close to zero for our demo purpose. Fig. 6.8 shows the graph of the function $f(t) = \sin^3(t) + \cos(3t)$ with t_5 marked by a solid red circle.

6.3.5 Mini project: Analysis of solution dependency on a parameter.

In this mini project you will revisit the system from Exercise 6.9 and use the resultant method to analyze how the real solutions of the system depend on the parameter a. Consider the polynomial system $p(x, y) = 0$, $q(x, y) = 0$ with

$$p(x, y) = x^2 - 2x - y = 0, \quad q_a(x, y) = a^2 + x^2 + y^2 - 2a\,y - 2x = 0.$$

Do the following:

Step 1. Using CAS assistance, construct the resultant $\text{Res}(p, q_a, x)$ and solve the equation $\text{Res}(p, q_a, x)$ for y. Determine the range of the parameter a for which the y-values are real.

Step 2. Solve the equation $p(x, y) = 0$ for x-values corresponding to the y-values found in Step 1. **Caution:** The range of the parameter values found in Step 1 must be reduced to ensure the existence of real x-values.

Step 3. Find the range of values of the parameter for which the system has two real solutions. Choose one of these parameter values, $a = a_0$, plot the curves $p(x, y) = 0$, $q_{a_0}(x, y) = 0$, and calculate the corresponding solutions of the system.

6.4 Implicitization of plane curves

Suppose a plane curve is defined in parametric form as $x = p(t)/r(t)$, $y = q(t)/r(t)$, where p, q, r are polynomials. We can rewrite these parametric equations as a polynomial system $x \cdot r(t) - p(t) = 0$, $y \cdot r(t) - q(t) = 0$ and apply the resultant method to eliminate the parameter t. The result is an implicit equation of the given curve. Here is a simple example.

Example 6.17. Find an implicit equation of the curve defined by the rational parametrization $x = (t^2 - 1)/(t^2 + 1)$, $y = 2t/(t^2 + 1)$.

Solution. We first rewrite the parametric equations as a polynomial system in the variable t with coefficients depending on x and y:

$$(1 - x)t^2 - x - 1 = 0, \quad yt^2 - 2t + y = 0.$$

The Sylvester matrix is then constructed as

$$S(x, y) = \begin{bmatrix} 1 - x & 0 & -x - 1 & 0 \\ 0 & 1 - x & 0 & -x - 1 \\ y & -2 & y & 0 \\ 0 & y & -2 & y \end{bmatrix}.$$

Using CAS assistance, the resultant is found to be $4x^2 + 4y^2 - 4$. Thus, an implicit equation of the given curve is $x^2 + y^2 - 1 = 0$. This is the unit circle centered at the origin. (This should be of no surprise. We found the rational parametrization of this circle in Chapter 4.)

We are not aware whether it is always possible to implement implicitization for a parametric equations of a curve defined by **trigonometric polynomials**.[5] However, in the next example we show a successful implicitization of one of such curves, an ellipse. (To simplify notation we introduced the coefficient 2 in the parametric equations of the curve.)

Example 6.18. Derive an implicit equation for the curve $x = 2a\cos(t)$, $y = 2b\sin(t)$.

---

[5]By this term we mean here a polynomial in x and y with the variables replaced with $\sin(t)$ and $\cos(t)$.

Solution. Using formulas (2.9) and the notation $z = \exp(it)$, we obtain after some elementary algebra

$$az^2 - xz + a = 0, \quad bz^2 - iyz - b = 0.$$

The Sylvester matrix for this system viewed as univariate in the variable z is

$$S(x, y) = \begin{bmatrix} a & -x & a & 0 \\ 0 & a & -x & a \\ b & -iy & -b & 0 \\ 0 & b & -iy & -b \end{bmatrix}.$$

With CAS assistance we find $\det S(x, y) = 4a^2b^2 - a^2y^2 - b^2x^2$, and the solvability condition $\det S(x, y) = 0$ yields the standard equation of the ellipse $x^2/(2a)^2 + y^2/(2b)^2 = 1$.

6.4.1 Lab 7: Implicitization of plane rational curves.

Problem formulation. Consider two families of curves defined by the parametric equations with two parameters:

$$\text{Curve 1:} \quad x = \frac{at^2 - 1}{t^2 + t + 1}, \quad y = \frac{bt^2 + t}{t^2 + t + 1}$$

$$\text{Curve 2:} \quad x = \frac{bt^2 - t + b}{t}, \quad y = \frac{at^2 + 2t - a}{t}$$

For each of the curves do the following.

Part 1. Use the resultant method to derive the corresponding implicit equation. The implicitization will show that the curves are conic sections.

Part 2. Identify the non-degenerate conic type that can be represented by the implicit equation found in Part 1.[6]

Part 3. Determine the parameter values for which the conic is degenerate.

Part 4. Pick parameters a, b such that the corresponding conic is not degenerate and plot the conic.

Part 5. Pick parameters a, b such that the corresponding conic is degenerate and plot the conic.

Suggested directions: The first task is well represented by numerous examples in this chapter. The result here is a certain polynomial with coefficients depending on the parameters a and b. Determine the coefficients $A(a, b)$, $B(a, b)$, $C(a, b)$ of x^2, xy and y^2 of this polynomial and compute the discriminant $d(a, b) = B^2 - 4AC$. This is the main tool for implementing the other tasks of the lab that require only basic algebraic skills and making plots using CAS assistance.

---

[6]Recall that the type of a conic defined by the equation $Ax^2 + Bxy + Cy^2 + Dx + Ey + F = 0$ depends on the discriminant $d = B^2 - 4AC$.

6.5 Glossary

- **Cramer's rule** – An explicit formula for the solution of a system of linear equations with as many equations as unknowns. It is valid whenever the system has a unique solution.

- **Reducible polynomial** with real coefficients – A non-constant polynomial that can be factored into a product of two non-constant polynomials with real coefficients.

- **Sylvester matrix of two univariate polynomials** – A certain $N \times N$ matrix constructed from the coefficients of the polynomials. Here N is the sum of the degrees of these polynomials.

- **Resultant of two univariate polynomials** – The determinant of Sylvester matrix of two univariate polynomials. It is equal to zero if and only if the polynomials have a common root.

- **Rational parametrization of a plane curve** – A representation of a curve by parametric equations of the form $x = p(t)/r(t)$, $y = q(t)/r(t)$, where p, q, r are polynomials. A curve that admits such a representation is called **rational**.

- **Implicitization of a plane curve** – A method to convert a rational parametric equation of a curve into an equivalent implicit form defined by a polynomial equation.

Part 2

Calculus and Numerics

7

Derivatives

Calculus is an elementary part of a field called mathematical analysis. This field also includes real analysis, complex analysis, differential equations, functional analysis, and some other branches of mathematics. Other mathematical and scientific fields, such as number theory, probability, and physics, are both using and developing methods of mathematical analysis.

Calculus was developed independently by I. Newton and G. Leibniz in the 17th century, but the rigorous foundations of calculus were only completed in the 19th century. This was mostly due to the work of A.-L. Cauchy and K. Weierstrass, with contributions from many other brilliant mathematicians along the way, such as L. Euler and B. Riemann.

Calculus provides tools to quantitatively investigate various processes of change, motion, and, generally, dependence of one quantity on one or more other quantities. It is commonly divided into two major branches: differential calculus and integral calculus. The two branches are linked by the remarkable fact that differentiation and integration are inverse operations. This fact is rigorously stated in the Fundamental Theorem of Calculus. The key fundamental concepts of calculus are functions, limits, derivatives, and integrals.

The derivative, the main concept of differential calculus, is a powerful tool for various mathematical tasks, like analyzing monotonicity and finding points of extrema (maxima and minima). Historically, the introduction of the derivative into mathematics was motivated by the problem of finding the slope of the tangent line at a given point on a curve. In applications, if a function $y = f(x)$ models the dependence of some quantity y on another quantity x, the derivative value $f'(x_0)$ gives the instantaneous rate of change of the quantity y when the quantity x equals x_0.

Example 7.1. Consider a thin rod of length l. Suppose that its shape is modeled by the interval $[0, l]$ on the x-axis. Let $m(x)$ be the mass of the rod segment $[0, x]$, $0 \le x \le l$, and $x_0 \in (0, l)$. Then $m'(x_0)$ is the **mass density** of the rod at the point x_0.

Example 7.2. Consider a point object (i.e., an object whose dimensions are ignored) moving along a straight line. The motion can be modeled by a point moving along the

x-axis starting from some initial position x_0 at time t_0. Let $d(t)$ be the position of the object at the time $t > t_0$. Then $d'(t)$ is the **velocity** of the object at the moment t.

The definite integral, the main concept of integral calculus, was historically motivated by the so called quadrature problem, which entails finding the area under the graph of a nonnegative function. In applications, if a function $y = f(x)$ models the instantaneous rate of change of a quantity y with respect to another quantity x on some interval $[a, b]$, then the definite integral of $f(x)$ over $[a, b]$ gives the accumulation of change (the net change) of y on this interval.

The examples above can be modified to illustrate the meaning of the definite integral in applications.[1]

Example 7.3. If $\rho(x)$ models the mass density of a thin rod of length l at a point x, the integral of $\rho(x)$ over the interval $[0, l]$ gives the **mass** of the rod.

Example 7.4. Suppose that a function v depending on time t models the velocity of an object moving along a straight line. If you choose the initial position of the object at time t_0 as the origin, the integral of $v(t)$ from $t = t_0$ to $t \geq t_0$ gives the **position** of the object at time t.

Performing the operations of differentiation and integration on a function reveals properties of the function that cannot be seen with the naked eye. These properties are important for both mathematical operations on functions and in applications, where functions are used as models of relationships between real life quantities. In this module we will briefly review some basic terms and facts about differential and integral calculus of functions of a single variable and illustrate several basic theorems. Many exercises require making CAS functions that implement certain tasks involving derivatives and integrals. Later in the module, we will focus on a few basic methods for approximating functions and their roots.

7.1 Review: Definitions, notation, and terminology

All functions in this module are assumed to be real-valued. Recall the formal definition of the **derivative** of a function f at a point c:

$$f'(c) = \lim_{\triangle x \to 0} \frac{f(c + \triangle x) - f(c)}{\triangle x}. \tag{7.1}$$

It is assumed here that the function is defined in a neighborhood of the point c.

A function is called **differentiable** at a point c if the derivative defined by (7.1) exists at c. For small $\triangle x$, the **difference quotient** on the right hand side of equation (7.1) can be used as an approximation of the derivative at the point c:

$$f'(c) \approx \frac{f(c + \triangle x) - f(c)}{\triangle x}. \tag{7.2}$$

If we change the notation $c + \triangle x$ in equation (7.2) to x and solve the equation for $f(x)$, we obtain a **local linear approximation of the function**. It is called the

---

[1] In the examples in this introduction, it is assumed by default that the derivative or the definite integral exists.

linearization of function near a point c.

$$f(x) \approx f(c) + f'(c)(x - c). \tag{7.3}$$

Geometrically, this equation means that near the point c the graph of a differentiable function f can be approximated by the straight line

$$y = f(c) + f'(c)(x - c).$$

Recall that this straight line is the tangent to the graph of the function f at the point $(c, f(c))$.

Exercise 7.5. Make a CAS function *fcn_and_tangent*(f, a, b, c, del) that takes

- a function f defined on the interval $[a, b]$ and differentiable in (a, b)
- a point $c \in (a, b)$
- a small float del^2

and returns

- a figure with the graph of f on $[a, b]$ and the tangent passing through the point $(c, f(c))$
- the relative error of the approximation of $f(c + del)$ by the linearization (7.3) at the point c

Suggestion: Use a CAS differentiation command to find the derivative of the input function or find the derivative manually and include it as an additional input argument of your function.

A function f that is differentiable at each point of an interval defines a new function on this interval, the derivative function f'. The graph of a differentiable function defined on some interval is a smooth curve over this interval. The derivative of f', typically denoted as f'', is the second derivative of the function f. Another notation for derivatives is $\frac{df}{dx}$ for the first derivative and $\frac{d^2 f}{dx^2}$ for the second one. For functions depending on the time variable, the notation $\dot{f}$ with the dot above the function name is often used in physics for the first derivative and $\ddot{f}$ with two dots above the function name for the second derivative.

Exercise 7.6. Make a CAS function *f_and_fprime*(**fcn, a, b**) that takes a differentiable function fcn on an interval $[a, b]$ and returns a figure with plots of the function and its derivative on the interval. Use a CAS command for finding the derivative function in your code.

The expression $df = f'(x)\triangle x$ is called the **differential of the function f at the point x**. Typically it is denoted as $df = f'(x)dx$. This gives a new interpretation to the notation $\frac{df}{dx}$ for the first derivative as the ratio of the differential of the function to the differential of the function argument.[3] Notice that the differential depends on two quantities: the value of the derivative $f'(x)$ and an increment $\triangle x$ of the argument x.

---

[2] Floats represent real numbers and are written with a decimal point dividing the integer and fractional parts.

[3] This interpretation is used, in particular, in solving so called separable ordinary differential equations (ODEs). Examples of simple separable ODEs will be used in some exercises later in this module.

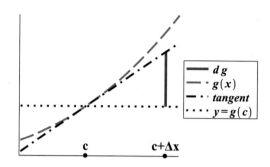

Figure 7.1. Differential dg is the length of the red segment. It approximates the change of the function g over the interval $[c, c + \triangle x]$.

Let g be a function differentiable in a neighborhood of the point c and $x = c + \triangle x$ a point in this neighborhood. Then equation (7.3) can be rewritten for the function g as $\triangle g \approx dg$, with $\triangle g = g(x) - g(c)$ being the change of g caused by adding an increment $\triangle x$ to $x = c$ and $dg = g'(c)dx$ the differential of the function g at the point c. This approximation property makes the differential useful in applications as shown in a simple example below. Fig. 7.1 gives a geometric interpretation of this property.

Example 7.7. Approximate the change in the radius of the base of a cone of height h needed to increase the cone volume by 10%.

Solution 1 (Using the differential of volume). We consider the volume of a cone with base radius r and height h as a function of the radius only:

$$V(r) = \pi r^2 h/3.$$

If the radius increases by dr, the corresponding change in volume can be approximated by the differential

$$dV = \frac{2}{3} \pi r h\, dr.$$

To get a 10% increase in volume, we equate the ratio $(dV)/V$ to 0.1 and solve for dr:

$$\frac{dV}{V} = \frac{2\,dr}{r} = 0.1 \Rightarrow dr = 0.05r.$$

Therefore, for a 10% growth in the volume of the cone, the radius should be increased by about 5%.

Solution 2 (Brute force solution). To obtain a 10% volume increase, the radius increment $\triangle r$ must satisfy the equation $V(r + \triangle r) = 1.1\, V(r)$.

⚓ **Check Your Understanding.** Show that this is a quadratic equation in $\triangle r$.

Solving the equation, we find $\triangle r = (\sqrt{1.1} - 1)r \approx 0.0488r$. Notice that the difference between the two solutions is about 0.12%.

Exercise 7.8. Use the differential to approximate the change in the radius of a ball needed to increase the volume of the ball by 12%.

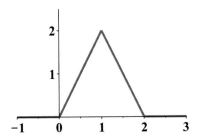

Figure 7.2. Continuous piecewise linear function.

A function is called **piecewise defined** if it is defined by two or more sub-functions. More precisely, the domain of a piecewise defined function consists of a finite number of subintervals, and on each of these subintervals the function is defined by a sub-function.[4] In approximation theory and its applications, piecewise polynomial functions called **splines** are used for interpolating complex functions or data. Some details of this technique are introduced later in this module.

Exercise 7.9. Consider the continuous piecewise linear function depicted in Fig. 7.2.

(a) Define and plot this function and its derivative in one figure.

(b) Make a CAS function **piecewise_linear_fcn(L)** that takes a list L of data points of a continuous piecewise linear function defined by these points and returns a figure with the plot of the function and its derivative. Assume that the points in the list L are sorted in nondecreasing order of their x-coordinates. **Suggestion:** Use a *for* loop defined by the number of segments in the graph of the function. To plot the function use a CAS command that plots a segment with given endpoints. Include a calculation of the slopes of the segments and plotting the derivative as a piecewise constant function in the body of the loop.

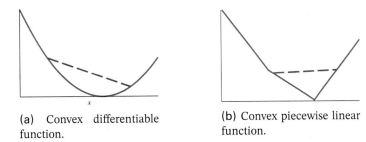

(a) Convex differentiable function.

(b) Convex piecewise linear function.

Figure 7.3. Convex functions.

---

[4]Notice that a piecewise linear function is completely defined by its values at the endpoints of the subintervals, and its derivative is a piecewise constant function.

7.2 Convexity of a univariate function

Geometrically, a function f is **convex** (also commonly called **concave up**) on an interval if for any two points on the function graph the segment connecting the points lies above or on the graph of f, see Fig. 7.3. This can be stated more formally as follows.

Definition 7.10. Let f be a function defined on some interval I. We say that f is convex on I if for any pair of points x_1, $x_2 \in I$ and any $\lambda \in [0, 1]$

$$f(\lambda x_1 + (1 - \lambda) x_2) \le \lambda f(x_1) + (1 - \lambda) f(x_2).$$

Notice that for $\lambda \in (0, 1)$, the left-hand side of this inequality is the value of f at a point inside the interval $[x_1, x_2]$ and the right-hand side is the y-value of the secant at the same point. Thus the definition is just an algebraic formalization of the intuitive, geometric description of convexity given above.

We now state two important facts about convex functions that will be useful in the next mini project.

Theorem. *A piecewise differentiable function*[5] *is convex if and only if its derivative is nondecreasing.*

As an example of applying this theorem, a visual inspection of the plots in Fig 7.3 suggests that the derivatives of the two depicted convex functions are nondecreasing.

Theorem. *If there is a point on the graph of a continuous piecewise linear convex function such that the slope of the segment on the left is negative and the slope of the segment on the right is positive, then the function has an absolute minimum at this point.*

☐ *Check Your Understanding.* (a) Sketch several piecewise linear convex functions to convince yourself of the above theorem. (b) Suppose that a piecewise linear convex function is constant on some subinterval in its domain. What can be concluded about the value of this constant compared with the other values of the function?

An applied problem in the next mini project involves the so called **geometric median**. In general, the geometric median of a discrete set of points in $\mathbb{R}^n$ is the point minimizing the sum of distances to the points of the set. The goal of the mini project is to find the geometric median of a discrete finite set of points on the line.

7.2.1 Mini project: Geometric median.

Problem formulation. Imagine that you are a postmaster for a district in some state. There are n towns along a straight road. Let x_j, $j = 0, \ldots, n - 1$, be the locations of the towns along the road arranged in ascending order. Your job is to determine the best location for a new post office that should be built in one of the towns. It is assumed that the best location is the one that minimizes the average distance between the post office and the towns. Notice that minimizing the average distance is equivalent to minimizing the sum of the distances from the post office to all towns.

---

[5]A piecewise defined function is called piecewise differentiable if (a) its constituent functions are differentiable on the corresponding open subintervals; (b) the one-sided derivatives exist at all subintervals endpoints.

Suggested directions.

Step 1. Temporarily relax the problem conditions and assume that the post office can be located at any point x on the interval $[x_0, x_{n-1}]$. Let $f(x)$ be the sum of distances from x to all towns (the objective function):

$$f(x) = \sum_{j=0}^{n-1} |x - x_j|.$$

Denote by f_k the restriction of this objective function to the interval $[x_{k-1}, x_k]$. For example, if $n = 5$ then

$$f_3(x) = (x - x_0) + (x - x_1) + (x - x_2) - (x - x_3) - (x - x_4),$$

for all $x \in [x_2, x_3]$. Derive the formula

$$f_k(x) = (2k - n)x + \sum_{j=k}^{n-1} x_j - \sum_{j=0}^{k-1} x_j. \tag{7.4}$$

Show that the piecewise linear objective function (7.4) is continuous and convex on the interval $[x_0, x_{n-1}]$.

Step 2. Analyze how the slopes of the linear pieces of f change signs for two cases:

- Case 1: n is odd, that is, $n = 2m + 1$.
- Case 2: n is even, that is, $n = 2m$.

For each case, plot the objective function for a concrete example with some small value of n.

Step 3. Based on your analysis in Step 2, write the symbolic optimal solution to the problem.

7.3 Some facts about functions and derivatives

7.3.1 Monotone Convergence Theorem (MCT).
Recall that a sequence in mathematics is defined as a function whose domain is the set of natural numbers. A sequence can be defined explicitly, such as $a_n = 0.5^n$, $n \geq 0$, or recursively, such as $a_0 = 1$, $a_1 = 1$, $a_n = a_{n-1} + a_{n-2}$, $n \geq 2$. (The latter is the famous Fibonacci sequence.) A sequence x_n, $n \geq 0$, is called **monotonic increasing** if $x_{k+1} \geq x_k$ for all k and **monotonic decreasing** if $x_{k+1} \leq x_k$ for all k. It is called **monotonic** in either of these two cases.

Theorem 7.11 (MCT). *Any monotonic and bounded sequence has a limit.*

Example 7.12. Use the MCT to show that each of the sequences below has a limit. (a) $a_n = 1 + 1/n$; (b) (optional) $b_n = (1 + 1/n)^n$.

Solution.

(a) Clearly the sequence $\{a_n\}_{n=1}^{\infty}$ is decreasing and bounded. (**Show this!**) It is easy to show that $\lim_{n \to \infty} a_n = 1$.

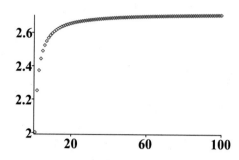

Figure 7.4. The first 100 terms of the sequence b_n, Example 7.12(b).

(b) Numerics (see Fig. 7.4 showing the first 100 terms of the sequence b_n) suggests that the sequence is increasing and bounded, but the rigorous proof of these assumptions of the MCT is challenging. (A sketch of the proof is given at the end of this subsection.) The sequence arises, for example, in the study of the compound interest model. Its limit is the remarkable mathematical constant $e \approx 2.718$, the base of the natural logarithm.

Some statements about sequences have their counterparts for functions. In particular, there is a theorem for functions similar to the MCT for sequences. One applied example of a monotonic and bounded function is the logistic function defined as $S(x) = 1/(1 + \exp(-x))$. This function is used in population dynamics and some algorithms of Machine Learning, a popular subfield of AI, where it is called the "sigmoid function".

Exercise 7.13.

(a) Show that the logistic function $S(x) = 1/(1 + \exp(-x))$ is bounded and strictly increasing.

(b) Find the limits of $S(x)$ as $x \to \pm\infty$.

(c) Plot the function and explain why it can be used for modeling probability.

Verification of MCT conditions for Example 7.12(b). To rigorously show that the sequence $b_n = (1 + \frac{1}{n})^n$ is bounded, let us use the binomial formula (Recall that $\binom{n}{k}$ are the binomial coefficients defined by the equation $\binom{n}{k} = \frac{n(n-1)\cdots(n-k+1)}{k!}$.)

$$(x + y)^n = \sum_{k=0}^{n} \binom{n}{k} x^k y^{n-k}.$$

We have

$$\left(1 + \frac{1}{n}\right)^n = 1 + \left(n \cdot \frac{1}{n} + \frac{n(n-1)}{2!}\frac{1}{n^2} + \frac{n(n-1)(n-2)}{3!}\frac{1}{n^3} + \cdots + \frac{1}{n^n}\right).$$

The right-hand side of this equation is less than $1 + \left(1 + \frac{1}{2} + \frac{1}{2^2} + \cdots + \frac{1}{2^{n-1}}\right)$. (**Show this!**)

$\boxed{\dot{\diamond}}$ **Check Your Understanding.** Prove that $1 + \left(1 + \frac{1}{2} + \frac{1}{2^2} + \cdots + \frac{1}{2^{n-1}}\right) < 3$.

Thus, the sequence is bounded.

To prove the monotonicity of the sequence, we will apply the binomial formula to two consecutive terms, b_n, b_{n+1}, and compare the kth terms in these expansions. For b_n, the kth term is

$$\frac{n(n-1)\cdots(n-(k-1))}{k!}\frac{1}{n^k} = \frac{(1-1/n)(1-2/n)\cdots(1-(k-1)/n)}{k!}.$$

For b_{n+1}, the kth term is

$$\frac{(n+1)n\cdots(n+1-(k-1))}{k!}\frac{1}{(n+1)^k} = \frac{\left(1-\frac{1}{n+1}\right)\left(1-\frac{2}{n+1}\right)\cdots\left(1-\frac{k-1}{n+1}\right)}{k!}.$$

The latter is larger, which proves the monotonicity of the sequence b_k.

7.3.2 Intermediate Value Theorem (IVT, Cauchy's Theorem).

Theorem 7.14 (IVT). *Let f be a continuous function on a closed interval $[a, b]$. Then for any number p between $f(a)$ and $f(b)$ there is a point $c \in (a, b)$ such that $f(c) = p$.*

In particular, if a continuous function takes values of different signs at the endpoints of an interval, the theorem guarantees that the function has at least one zero inside the interval. The **bisection method**, a simple method for finding zeros of continuous functions, is based on this fact.

Description of the basic bisection algorithm. Let f be a continuous function on $[a, b]$ such that $f(a) \cdot f(b) < 0$.

Step 1. Find the midpoint $c = (a + b)/2$ of the interval $[a, b]$ and calculate $f(c)$.

Step 2. If $f(a) \cdot f(c) < 0$, set $b = c$; otherwise set $a = c$.

Step 3. Repeat Steps 1 and 2 until a certain stopping criterion is satisfied.

Possible choices for the stopping criterion could be a fixed number of iterations, a sufficiently small length of the current interval, or a sufficiently small value of the function at the midpoint of the current interval.

The next example shows a few iterations of the basic bisection method. It is very intuitive and reliable for an interval with a single root, but rather slow.

Example 7.15.

(a) Show that the function $f(x) = \cos(x) - x$ has one zero on the interval $[0, 1]$.

(b) Implement four steps of the bisection method.

Solution.

(a) Clearly $f(x)$ is continuous for all $x \in [0, 1]$. Since $f(0) = 1 > 0$ and $f(1) = \cos(1) - 1 < 0$, the function has a zero in $[0, 1]$. The cosine function is decreasing and the function $y = x$ is increasing on the interval. Therefore, the function f is strictly decreasing.[6] We conclude that f has only one zero on the interval $[0, 1]$.

---

[6] Recall that in general the sign of the derivative of a differentiable function tells you whether the function is increasing or decreasing.

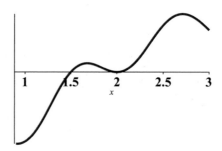

Figure 7.5. Basic bisection algorithm for this function stops on Step 2.

(b) The results of using the algorithm are given in the table below:

a	b	c	$f(c)$
0.00000	1.00000	0.50000	0.37758
0.50000	1.00000	0.75000	-0.01831
0.50000	0.75000	0.62500	0.08533
0.62500	0.75000	0.68750	0.03388

Based on these calculations, we conclude that the zero of f is in the interval $[0.68750, 0.7500]$. The "exact" value found using a high precision root-finding command is 0.73905.

Remark 7.16. If the initial interval contains more than one root, the basic bisection algorithm described above can fail. For example, the function in Fig. 7.5 has different signs at the endpoints of the interval $[1, 3]$, but the basic bisection algorithm stops on Step 2. (**Explain why.**)

Exercise 7.17. Make a CAS function *my_bisection*(**fcn, a, b, eps**) that takes a function fcn, endpoints of an interval with a single root of the function, and a small float *eps*, and returns an approximation of the root. Use the following stopping criterion: repeat the iterations until the value of the function at the midpoint of the current interval becomes less than *eps*. Test the function on an example of your choice.

7.3.3 Mean Value Theorem (MVT, Lagrange Theorem).

Theorem 7.18 (MVT). *If a function f is continuous on a closed interval $[a, b]$ and differentiable in (a, b), there exists a point $c \in (a, b)$ such that*

$$\frac{f(b) - f(a)}{b - a} = f'(c).$$

Fig. 7.6 shows a geometric interpretation of this theorem.

The theorem is used in proving many important mathematical statements. A simple example is the proof of the following fact: If a function f satisfies the conditions of the MVT on $[a, b]$ and $f'(x) = 0$ for all $x \in (a, b)$, then the function is constant. (**Prove this!**)

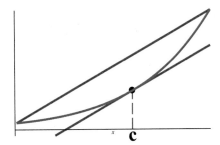

Figure 7.6. Geometric meaning of the MVT: The slope of the tangent to the curve $y = f(x)$ at $x = c$ is equal to the slope of the secant connecting the endpoints of the curve.

Exercise 7.19. Make a CAS function *my_MVT*(**fcn, a, b**) that takes a function *fcn* that is continuous on an interval $[a, b]$ and differentiable in (a, b), and returns a plot similar to the one in Fig. 7.6. Use a CAS root-finding command in the function code.

In the next mini project we use the derivative as the main tool for analyzing a function with many interesting properties.

7.3.4 Mini project: Analysis of a function with many characteristic features. Consider the function

$$f(x) = \frac{x}{x+1} - \cos\left(\frac{40}{x}\right).$$

Step 1. Find the domain of the function. Calculate the limits of the function as $x \to \pm\infty$.

Step 2. Plot the function. Your graph should include points of discontinuity and illustrate the behavior of the function as $x \to \pm\infty$ reasonably well. Since the function is unbounded, in order to make a nice picture, restrict the range of the y-values in the plotting command.

Step 3. Find the smallest negative and the largest positive x-intercepts of the function.

Step 4. If possible, determine how many x-intercepts the graph of this function has in the interval $(0, 10)$. If not, explain why.

Step 5. Find the largest local minimum on the interval $(-\infty, -1)$.

Exercise 7.20. Consider the function

$$f(x) = \begin{cases} 3 & \text{if } x = 0 \\ x/(x+1) - \cos(40/x) & \text{otherwise.} \end{cases}$$

(a) Plot the function on the interval $[-0.05, 0.05]$.

(b) Note f has a maximum at $x = 0$. If a differentiable function has a local maximum at a point, the derivative of the function changes sign from positive to negative when x passes through this point from left to right. Is there a neighborhood $(-\varepsilon, \varepsilon)$ of $x = 0$ such that on the left half of the interval $f'(x) > 0$ and on the right half of the interval $f'(x) < 0$? Explain.

7.4 Lab 8: Constructing a square circumscribed about ellipse

Closed plane curves, like ellipses, can be defined implicitly as $f(x, y) = 0$ or parametrically as $x = x(t), y = y(t), t_1 \le t \le t_2$. A curve defined parametrically is called **smooth** if the functions $x(t), y(t)$ have continuous first derivatives and $x'(t)^2 + y'(t)^2 \ne 0$ for $t_1 \le t \le t_2$. This is equivalent to saying that a simple, smooth, closed curve has a continuously varying tangent vector $\mathbf{s}(t) = [x'(t)\ y'(t)]$.[7] Notice that the slope of the tangent to such a curve, $m = (y'(t))/(x'(t))$, varies continuously on subintervals of $[t_1, t_2]$ where $x'(t)$ is nonzero.

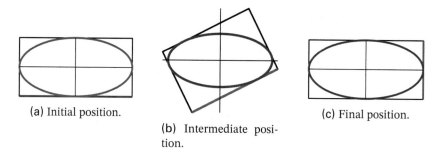

(a) Initial position.

(b) Intermediate position.

(c) Final position.

Figure 7.7. Snapshots of the moving rectangle circumscribed about an ellipse.

Problem formulation. It is known that for any ellipse $x = a\cos(t), y = b\sin(t), t \in [0, 2\pi)$, there exists a square circumscribed about it. Derive the symbolic equations for the sides of the square. Choose a specific ellipse and use CAS assistance to plot the ellipse and the circumscribed square in one figure.

About existence of a square circumscribed about an ellipse. A sketch of the proof of existence of a circumscribed square about any ellipse is given below. It relies on the IVT and will be helpful for developing some intuition about the problem. A more general statement about the existence of a circumscribed square about any simple closed curve is also valid.

Proof. Suppose the ellipse $x^2/a^2 + y^2/b^2 = 1$ is initially enclosed in a rectangle of length $l_0 = 2a$ and width $w_0 = 2b$ (Fig. 7.7a). Without loss of generality we can assume that $a > b$. Imagine that *the rectangle is moving counterclockwise in such a manner that its length and height are changing, but the rectangle remains circumscribed about the ellipse.* Assume that the initial base of the rectangle (the red side in Fig. 7.7a) preserves its color in this process. Eventually, it becomes vertical (Fig. 7.7c).

Fig. 7.7b shows a generic intermediate position of the moving and changing rectangle and Fig. 7.7c depicts the final position of the ellipse. When the point of tangency A of this side traverses the lower right quarter of the ellipse, the value of the slope m of the tangent at this point varies continuously and monotonically in the interval $[0, \infty)$. The length $l(m)$ and the width $w = w(m)$ of the changing rectangle are also continuous functions of m.

---

[7]A closed plane curve is called simple if it does not intersect itself.

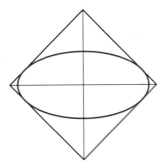

Figure 7.8. The square circumscribed about ellipse.

Consider the function $f(m) = l(m) - w(m)$. It is continuous on $[0, \infty)$, positive at $m = 0$, and negative for sufficiently large m. By the IVT and monotonicity of m, the function f has a unique zero m^* on a sufficiently large interval $[0, M]$. For $m = m^*$, we have $l(m^*) = w(m^*)$, that is, the rectangle is a square.   □

The challenge of the lab is to actually construct the circumscribed square.

Suggested solution steps and directions.

Part 1. Derive a **reparametrization** of the ellipse using the **slope of the tangent**

$$m = \frac{y'(t)}{x'(t)} = -\frac{b}{a} \cot(t)$$

as a new parameter. Note that the points $(\pm 1, 0)$ with vertical tangents will not be covered by this parametrization. (**Why?**)

Directions: Some trigonometry is required to express $\sin(t)$ and $\cos(t)$ in the standard parametrization of the ellipse as functions of the new parameter m. Specifically, we can express these two trig functions in terms of $\cot(t) = -(a\,m)/b$. For different quadrants, geometric considerations determine the sign choices for $\sin(t)$ and $\cos(t)$. For instance, for the lower right quarter of the ellipse, sine is negative and cosine is positive. Thus the functions $x = x(m), y = y(m)$ of the new parametric equations of the ellipse are piecewise-defined.

Part 2. Find the equations of the sides of a moving rectangle circumscribed about the ellipse.

Directions: Let A be a point $(x(m), y(m))$ of tangency of the red side of the moving rectangle (Fig 7.7b), where m is the slope of this tangent to the ellipse. Find the coordinates of the other three points of tangency[8] and the slopes of the three tangents passing through these points. Then, since a point and slope uniquely define the equation of any line, the equations of all sides of the rectangle can be found in terms of m.

Part 3. Find the equation of the function $f(m) = l(m) - w(m)$ and the parameter value that defines the circumscribed square.

---

[8]all of them are functions of the same value m.

Directions: Use the formula for the distance between a point and line or find the three vertices of the moving rectangle and calculate the lengths of its sides as the distances between consecutive pairs of these vertices.

The output of this part is the exact value of the slope m^* of the red side of the circumscribed square. If you do this part correctly, you will get a surprise: **The slope m^* does not depend on the semi-axes** a, b **of the ellipse!**

Part 4. Consider the ellipse with $a = 2$, $b = 1$. Write the equations of the four sides of the circumscribed square and plot the ellipse and the four sides of the square in one figure. Your figure will be similar to the one in Fig. 7.8.

Exercise 7.21. Find a square circumscribed about a rectangle defined by its vertices. For simplicity, assume that the rectangle is symmetric with respect to the coordinate axes. It goes without saying that the given rectangle is not a square.

7.5 Glossary

- **Derivative of a function at a point**:

$$f'(c) = \lim_{\triangle x \to c} \frac{f(c + \triangle x) - f(c)}{\triangle x}$$

- **Tangent line** to a plane curve at a given point – Informally, the straight line that "just touches" the curve at the given point. For a curve $y = f(x)$ with f differentiable at $x = a$, the tangent line at the point $(a, f(a))$ is defined by the equation $y = f(a) + f'(a)(x - a)$.

- **Difference quotient**:

$$\frac{f(c + \triangle x) - f(c)}{\triangle x}$$

- **Differential**:

$$df = f'(x)dx$$

- **Convex** function on an interval I – A function f whose graph on the interval I satisfies the following condition: for any two points on the graph, the segment connecting the points lies above or on the graph. Algebraically, this means that for any x_1, $x_2 \in I$ and any $\lambda \in [0, 1]$ the following inequality holds:

$$f(\lambda x_1 + (1 - \lambda) x_2) \le \lambda f(x_1) + (1 - \lambda)f(x_2).$$

8

Definite Integrals

The concept of a definite integral was historically motivated by the problem of finding the area under the graph of a nonnegative function (the quadrature problem). In fact, cases of finding simple areas and volumes by breaking them up into a larger and larger number of easily computed parts were even seen in ancient Greece and China. However, the systematic formal approach to integration started in the 17th century with the works of Newton and Leibniz. Since then many integration techniques have been developed and used in various applications.

After reviewing some basic facts from integration theory, we show by examples how the theory works for solving various problems. In addition, this chapter includes three mini projects and two labs that all require the use of various integration techniques. Problems in two of the mini projects are based on classic applications of integration in mechanics (projectile motion) and economics (Lorenz curves and the Gini index). The third mini project is on the derivation of a nice looking formula for the area of a tilted ellipse. The mathematical tool suggested for this derivation is the so called **shear transformation (shear mapping) of the plane**. It is known that the area of a region in the plane is unchanged under shear mappings. With an appropriate shear transformation, the equation of a tilted ellipse can be linked with the standard equation of an ellipse having the same area.[1]

Lab 9 involves calculating the volume of a body of revolution. Although such a calculation is a standard topic in univariate calculus, the problem in the lab is a simplified, educational version of a real industrial problem solved by a WCSU student[2] for a Connecticut manufacturing company. A computer implementation of this engineering problem is now part of the company's automated design. In Lab 10 a model of the motion of a skydiver that accounts for air resistance will be derived using Newton's Second law, solved, and analyzed.

---

[1]The derivation of the formula for the area of an ellipse given by the standard equation is a canonical textbook integration example.

[2]This former student is the second author of this book.

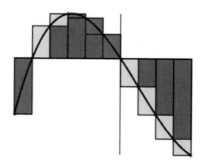

Figure 8.1. Visualization of the left and right Riemann sums. The darker color indicates where the rectangles representing different Riemann sums overlap.

8.1 Review: Some basic concepts and facts of univariate integral calculus

The rigorous definition of the definite integral is rather involved and is not explicitly stated in this chapter. However, we recommend to review this definition and relevant terms using any standard calculus text.

Recall that every continuous (or bounded, piecewise continuous) function on a finite interval is integrable, that is, the definite integral of the function over this interval exists. Riemann sums are approximations of definite integrals by finite sums. Fig. 8.1 shows the left and right Riemann sums with a uniform partition of the interval of integration. The darker color indicates where the purple and green rectangles overlap.

💡 ***Check Your Understanding.*** Identify by color the left and right Riemann sums in Fig. 8.1. Explain why there are two pure purple rectangles and two pure green rectangles, that is, there are no dark parts in these four rectangles.

Exercise 8.1. Make a CAS function ***riemann_sum*(f, a, b, n, p)** that takes five inputs:

- endpoints a, b of the interval of integration $[a, b]$ with $a < b$

- function f continuous on the interval $[a, b]$

- number n of subintervals in the Riemann sum

- parameter p taking one of the values 0, 1, or 2

and returns the

- left Riemann sum if $p = 0$

- midpoint Riemann sum if $p = 1$

- right Riemann sum if $p = 2$

Test the function on an example of your choice.
Suggestion: Use an *if-elif-else* conditional to encode three possible calculation options depending on the value of p.

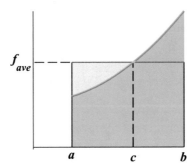

Figure 8.2. Illustration of the Mean Value Theorem for definite integrals. The area under the curve $y = f(x)$ equals the area of the rectangle with base $b - a$ and height $f_{ave} = f(c)$.

📖 Integration theory was first rigorously formalized based on the concept of a limit by Bernhard Riemann. The definite integral that you know is called the Riemann integral. In advanced analysis courses other types of integrals are defined and studied.

The **Fundamental Theorem of Calculus (FTC)** is at the heart of integral calculus. Informally, the theorem states that differentiation and integration are inverse operations, like multiplication and division or addition and subtraction. The theorem has two parts.

Theorem 8.2 (Fundamental Theorem of Calculus).

(1) *Let f be a continuous function on the interval $[a, b]$. Then the function F defined by*

$$F(x) = \int_a^x f(t)\,dt$$

is an antiderivative of f, that is, $F'(x) = f(x)$ on $[a, b]$.

(2) *Let f be a function defined on the interval $[a, b]$ and Φ some antiderivative of f on this interval. Then*

$$\int_a^b f(x)\,dx = \Phi(b) - \Phi(a). \tag{8.1}$$

💡 **Check Your Understanding.** It does not matter which antiderivative Φ of the function f is used in Part 2 of the FTC. Why?

Recall that the mean (average) value of a function f over an interval $[a, b]$ is defined by the formula

$$f_{ave} = \frac{1}{b - a} \int_a^b f(t)\,dt.$$

The Mean Value Theorem for integrals states that a continuous function on a closed interval attains its mean value at some point in the interval.

Theorem 8.3 (Mean value theorem for integrals). *Let f be a continuous function on the interval $[a, b]$. Then there is a point $c \in (a, b)$ where the function attains its mean value, that is*

$$f(c) = \frac{1}{b-a} \int_a^b f(t) \, dt. \tag{8.2}$$

Exercise 8.4.

(a) What is the geometric meaning of the formula (8.2) if $f(x) \geq 0$ on $[a, b]$?

(b) Make a CAS function **mean_value(f, a, b)** that takes the endpoints of a closed interval and a continuous function defined on this interval and returns a figure with the graph of the function, the line $y = f_{ave}$, and a point c such that $f(c) = f_{ave}$. Use built-in CAS functions for integration and root-finding in this exercise. Your figure will be similar to Fig. 8.2.

(c) Test the function made in part (b) on the following applied problem: Breathing is cyclic and a full respiratory cycle takes about 5 seconds. To model the amount of air present in the lungs we can use the function $f(t) = (1/2) \sin (2\pi t/5)$. Find the average volume of inhaled air in the lungs in one respiratory cycle and a moment when the average volume is attained.

Answer: $f_{ave} = 1/\pi \approx 0.318$ is attained at $t \approx 1.089$.

8.2 Area of a region bounded by a simple closed curve

Let $x = x(t), y = y(t)$, $t_1 \leq t \leq t_2$, be a parametric equation of a simple, smooth, closed curve. If a point $(x(t), y(t))$ moves counterclockwise when the parameter t changes from $t = t_1$ to $t = t_2$, the area enclosed by the curve can be calculated by the integral

$$A = \frac{1}{2} \int_{t_1}^{t_2} \left(x(t)y'(t) - y(t)x'(t) \right) dt. \tag{8.3}$$

The derivation of this formula involves tools of multivariate calculus, but its use is within the scope of this chapter since it requires calculating the integral of a function of a single variable.

Suppose a parametric curve can be also defined by two explicit functions $f_+(x)$ and $f_-(x)$ on the interval $[a = \min x(t), b = \max x(t)]$. The "top" function $f_+(x)$ bounds the enclosed region from above and the "bottom" function $f_-(x)$ bounds the region from below. Recall that in this case the area enclosed by the curve can be found by the integral

$$A = \int_a^b \left(f_+(x) - f_-(x) \right) dx. \tag{8.4}$$

Note that the formula (8.3) is more general since it works for parametric curves which do not admit such explicit representation by "top" and "bottom" functions.

🔅 **Check Your Understanding.** Make a sketch of a simple closed curve that is **not** a union of explicitly defined top and bottom parts.

Example 8.5. Find the area A enclosed by the ellipse

$$x = a \cos(t), y = b \sin(t), 0 \leq t < 2\pi.$$

Solution. From the implicit equation of an ellipse, we find $f(x) = \pm b\sqrt{1 - x^2/a^2}$. The curve is symmetric with respect to the x-axis, so we can find half of the area and double the result.[3] We obtain

$$A = 2b \int_{-a}^{a} \sqrt{1 - x^2/a^2} \, dx.$$

Using the change of variables $x = a\cos(t)$ makes the integrand more manageable:

$$A = 2ab \int_{\pi}^{0} \sin(t) \cdot (-sin(t)) \, dt = ab \int_{0}^{\pi} \left(1 - \cos(2t)\right) dt.$$

Here, the identity $\sin^2(t) = \left(1 - \cos(2t)\right)/2$ has been used. Calculating the integral yields the result $A = \pi ab$.

⚗ **Check Your Understanding.** Solve Example 8.5 using formula (8.3) to derive the same answer.

Example 8.6. Consider Fig. 8.3. The depicted region is bounded by a simple closed curve defined by the parametric equations $x(t) = 1 - t^2$, $y(t) = t - t^3$, $-1 \le t < 1$. Find the vertical line that halves the area of the region.

Solution 1 (Working with parametric equations of the curve). By plotting parts of the curve, observe that as the parameter t increases, the point traverses the curve counterclockwise starting from the origin. Since the curve is symmetric with respect to the x-axis, we can just solve the problem for the upper half of the region bounded by the curve. The area of this part can be found as

$$A_0 = \frac{1}{2} \int_{0}^{1} \left(x(t)y'(t) - x'(t)y(t)\right) dt.$$

Upon calculating this integral we obtain $A_0 = 4/15$. Now we introduce the integral with a variable upper limit as

$$A(s) = \frac{1}{2} \int_{0}^{s} \left(x(t)y'(t) - x'(t)y(t)\right) dt$$

and use CAS assistance to find the value of s such that $A(s) = 2/15$. CAS returns the value $s = 0.643138782$. The corresponding x-value is $x(s) \approx 0.586372507$. Thus the line $x = 0.586372507$ halves the area of the region (see Fig. 8.3).

Solution 2 (Working with explicit top part of the given curve). We solve the equation $x(t) = x$ for t to obtain $t_{\pm}(x) = \pm(1 - x)^{0.5}$. The explicit equation of the upper part of the curve is

$$top(x) = t_{+}(x) - (t_{+}(x))^3 = (1 - x)^{0.5} - (1 - x)^{1.5}.$$

Again, using symmetry of the bounding curve with respect to the x-axis, we find half of the enclosed area to be the same:

$$\int_{0}^{1} (1 - x)^{0.5} - (1 - x)^{1.5} \, dx = 4/15.$$

---

[3]Or find one quarter of the area and quadruple the result.

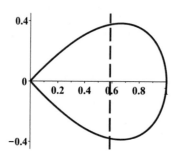

Figure 8.3. The region defined in Example 8.6 divided by the vertical line into subregions of equal areas.

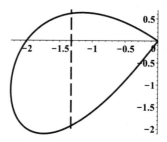

Figure 8.4. Halving area of a region non-symmetric with respect to x-axis, Exercise 8.7(b).

Now solving the equation

$$\int_0^c (1-x)^{0.5} - (1-x)^{1.5} \, d x = 2/15$$

for c, we obtain the same answer as before.

Exercise 8.7.

(a) Consider the parametric curve

$$x(t) = t^2 - 4, \; y(t) = t - t^3, \; -2 \le t \le 2.$$

A plot of the curve shows two parts of the area bounded by it – a curvilinear triangle and a loop. Ignore the loop. Find the vertical line that halves the area of the curvilinear triangle. Plot the curve and this vertical line in one figure.

(b) (Challenge!) Consider the parametric curve

$$x = t^2 - t - 2, \; y = t^3 - t^2 - 2t, \; -1 \le t \le 1.$$

Find the vertical line that halves the area bounded by this curve. Plot the curve and the line in one figure. Your figure will be similar to Fig. 8.4.

(a)   (b)

Figure 8.5. Smiling face (a) before and (b) after a shear transformation with shear factor $k = 2$.

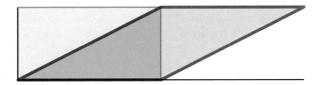

Figure 8.6. The image of a rectangle after the shear transformation. Both figures have the same area.

8.2.1 Mini project: Area of a tilted ellipse.

In this project you will construct and use a so called shear transformation (shear mapping) in the horizontal direction to derive the general formula for computing the area of a tilted ellipse. A horizontal shear transformation is defined by the matrix

$$S = \begin{bmatrix} 1 & k \\ 0 & 1 \end{bmatrix},$$

where k is a parameter called the **shear factor**. The transformation maps the radius-vector of a point (x, y) to the radius-vector of the point $(x + ky, y)$, that is

$$S \cdot \begin{bmatrix} x \\ y \end{bmatrix} = \begin{bmatrix} x + ky \\ y \end{bmatrix}.$$

As an example, the original shape of a smiling face and its image under the shear transformation S with shear factor $k = 2$ are shown in Fig. 8.5.

Recall that rigid transformations of the plane do not change the size or shape of figures. On the other hand, **shear mappings do change the shape of figures, but preserve their areas**. The image of the rectangle $D = \{(x, y) | 0 \le x \le 2, 0 \le y \le 1\}$ under the shear transformation S is depicted in Fig. 8.6. Clearly, the rectangle and its image (the parallelogram) have the same area.

Problem formulation.

Step 1. Find the general formula for the area of an ellipse defined by the equation[4] $A x^2 + 2 B x y + C y^2 = 1$, $B^2 - A C < 0$.

Suggested directions: Apply the completing the square technique to the first two terms in the left-hand side of the given equation. Then rewrite the result in the form

$$\frac{(x + k y)^2}{a^2} + \frac{y^2}{b^2} = 1. \tag{8.5}$$

---

[4]To simplify algebraic manipulations, we included the factor 2 in the second term of the equation.

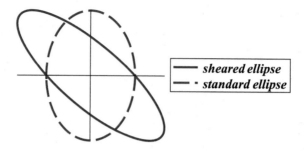

Figure 8.7. A standard ellipse and its image under a shear transformation have the same area.

The parameters k, a, b in equation (8.5) depend on the coefficients A, B, C of the given equation. Next, use the relation between the areas of the given ellipse defined by this equation and the ellipse defined by the standard equation with parameters a, b. This will lead you to the required formula.

Step 2. Apply the formula derived in Step 1 to the ellipse with parameters $A = 1/4$, $B = 2/5$, $C = 1$. Plot the given tilted ellipse and the corresponding standard ellipse in one figure. Your figure will be similar to Fig. 8.7.

📖 The determinant of any shear transformation is one. It is known that any linear transformation of either the plane or three-dimensional space defined by a matrix with determinant one does not change the area of figures or the volume of bodies.

8.2.2 Mini project: Lorenz curve and Gini index[5]. A **Lorenz curve** is given by the equation $y = L(x)$, where the independent variable x, $0 \le x \le 1$, represents the lowest fraction of a society's population in terms of wealth and $y(x)$, $0 \le y \le 1$, is the fraction of the total wealth that is owned by that fraction of society. For example, the Lorenz curve in Fig. 8.8 shows that $L(0.5) = 0.2$, which means that the lowest fraction 0.5 (50%) of the society owns 0.2 (20%) of the wealth. The figure also shows the line $y = x$ called the **line of perfect equality**.

Problem formulation. Do the following.

Step 1. Do the following.

(a) Explain why the line $y = x$ in Fig. 8.8 is given the name "line of perfect equality".

(b) Explain why any Lorenz curve must satisfy the conditions

$$L(0) = 0, \ L(1) = 1, \ L'(x) \ge 0 \text{ on } [0, 1]. \tag{8.6}$$

(c) Use CAS assistance to plot the Lorenz curves $L(x) = x^p$ with $p = 1.1, 2, 3, 4$ in one figure. Which of these values of p corresponds to the most equitable distribution of wealth? Which value of p corresponds to the least equitable distribution of wealth? Explain.

---

[5]This mini project is a slightly modified version of an exercise from [14].

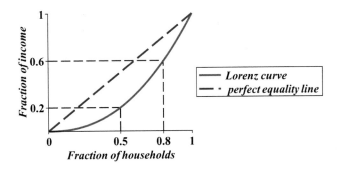

Figure 8.8. Lorenz curve and the line of perfect equality.

(d) The information on the Lorenz curve is often summarized in a single measure called the **Gini index**, which is defined as follows. Let A be the area of the region between $y = x$ and $y = L(x)$ and let B be the area of the region between $y = L(x)$ and the x-axis. The Gini index is defined as $G = A/(A + B)$. Show that

$$G = 1 - 2 \int_0^1 L(x) d x.$$

Step 2. Make a CAS function with no inputs that computes the Gini index $G(p)$ for the Lorenz curves $L(x) = x^p$, plots the function $f = G(p)$, $1 < p \le 5$, and returns the range $[c, d]$ of the function f. Which endpoints of the range correspond to the least and most equitable distribution of wealth?

Step 3. Make a CAS function **gini_index(f)** that takes a function f with nonnegative derivative on the interval $[0, 1]$, verifies the conditions $f(0) = 0$ and $f(1) = 1$, and returns the Gini index for the given Lorenz function and a plot similar to Fig. 8.8 if both conditions are true. Otherwise, the function returns the string "The given function does not define a Lorenz curve." Test your code on the function $L(x) = 5x^2/6 + x/6$.

Answer: The Gini index is 5/18.

Suggestion: Use an *if-else* conditional statement with a compound condition for testing if $f(0) = 0$ and $f(1) = 1$.

8.3 Lab 9: Submergence depth of a body of revolution in equilibrium

Relevant physics background. Archimedes' Principle states that the upward force exerted on an object immersed in a fluid is equal to the weight of the fluid the object displaces. This upward force is called the **buoyancy** or buoyant force. An object is in equilibrium (at rest) if the weight of the object is equal to the buoyant force (see Fig. 8.9.)

Notation.

- g – gravitational constant
- M – body mass

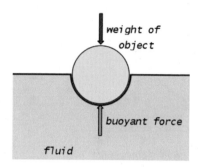

Figure 8.9. Illustration of the buoyancy principle.

- V – body volume

- ρ – mean body mass density, $\rho = M/V$

- W – weight of the floating body, $W = \rho\, g\, V$

- ρ_0 – mass density of the fluid

- V_0 – volume of the displaced fluid

- M_0 – mass of the displaced fluid, $M_0 = \rho_0\, V_0$

Archimedes' Principle implies that if a body immersed in a fluid is in equilibrium then

$$W = g\, M_0.$$

This equation can be rewritten as

$$V_0 = \frac{\rho}{\rho_0} V. \tag{8.7}$$

Problem formulation. Consider a hollow body bounded by an ellipsoid of revolution $x^2/a^2 + y^2/b^2 + z^2/b^2 = 1$. Let ρ be the mean mass density of the body and assume that the body is partially submerged in a fluid with mass density ρ_0. Determine the submerged depth of the body.

Plan and directions.

Part 1. Find the volume of the body. This can be done manually using the **disk method** or with CAS assistance.

Part 2. Calculate the volume V_0 of the displaced liquid as a function of the submergence depth h. This volume is the same as the volume of the submerged part of the body.

Directions: It is technically convenient to assume that the position of the body is fixed with the ellipsoid center at the origin and the initial fluid level at $z = -b$. Then we can picture submersion as the rise of the fluid level. Possible values for the submersion depth h are in the interval $[-b, b]$. The submerged part for some negative h might look like the bowl in Fig. 8.10. A horizontal plane $z = h$, $-b < h < b$, crosses the ellipsoid along an ellipse whose semi-axes are functions of h. The **slicing method** would work nicely for finding $V_0(h)$.

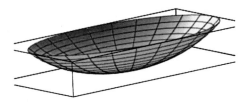

Figure 8.10. Possible shape of submerged part of the ellipsoid.

Part 3. Find the submergence depth. To simplify derivations, suppose that the mass density of the liquid ρ_0 equals one. Write equation (8.7) with $V_0 = V_0(h)$. The result is a cubic equation in h involving the parameter ρ. Find the submersion depth for the mean mass densities of the ellipsoid $\rho = 0.3, 0.6, 0.9$.

Directions: It is convenient (but not required) to introduce a new variable $y = h/b$. Solve the resulting cubic equation with the given mean mass density values of the body.

Part 4. Make a CAS function that takes parameters a, b of the ellipsoid of revolution about x-axis, the mean mass density ρ of the body, and the mass density ρ_0 of the liquid, and returns the submersion depth of the body in the liquid.

Part 5. Test your solution for $a = 2$, $b = 1$, $\rho = 0.5$, $\rho_0 = 1$. The answer for this test problem is $h = 0$. This means that for the given parameters exactly half of the ellipsoid is submerged.

8.4 Solving some ordinary differential equations

In the lab and mini project in this section, the reader will encounter equations that involve first and second derivatives of unknown univariate functions. These are particular types of so called **ordinary differential equations (ODEs)**. An ODE can include an unknown function f, its derivatives, and known functions of the argument of f. The highest derivative present in the equation is called the **order** of the ODE. The most general form of a **second order ODE** can be written as

$$F(x, f, f', f'') = 0. \tag{8.8}$$

If f'' is not present in the left-hand side, then equation (8.8) is a **first order ODE**. There is no general method for solving ODEs. In turn, various solution techniques have been developed over the years for certain classes of ODEs. In the next subsection we show a simple method for solving a special type of ODE called *separable*.[6] The method will be used in the next lab on finding the speed of a skydiver. The last project in this chapter on projectile motion involves solving two elementary second order ODEs that have the form $f'' = g(t)$, where $g(t)$ is a known function. The function f can be found by direct integration.

---

[6]We will not discuss more delicate issues, such as the existence or uniqueness of solutions. These are standard topics in courses on differential equations.

8.4.1 Solving separable ODE. Finding solutions to any separable ODE is a simple procedure that we will demonstrate in the next example.

Example 8.8. Find the solution to the first order differential equation $y' = -2xy$ satisfying the initial condition $y(0) = 2$.

This is an example of a so called **initial value problem, IVP.** A solution to the IVP is a function f that satisfies both the equation and the initial condition, that is, $f'(x) = -2x f(x)$ and $f(0) = 2$.

Solution. Rewrite the ODE as

$$\frac{dy}{dx} = -2xy. \tag{8.9}$$

Assuming $y \neq 0$, divide both sides of equation (8.9) by y and multiply by dx to obtain[7]

$$\frac{dy}{y} = -2x\,dx. \tag{8.10}$$

At this point we have successfully separated the variables x and y using just simple algebraic manipulations! Equation (8.10) tells us that the differential of some function of y whose derivative is $1/y$ equals the differential of some function of x with derivative $-2x$. We can easily guess what these functions are:

$$\frac{dy}{y} = d\left(\ln(|y|) + c_1\right), \quad -2x\,dx = d(-x^2 + c_2).$$

In general, we do not have to guess. After separating the variables, we can simply integrate both sides of the resulting equation to obtain the general solution to the differential equation. (Notice that the two arbitrary constants of integration on the left- and right-hand sides can be combined into one.) The general solution of a separable ODE often has an implicit form.

For our problem, the general solution is $\ln |y| = -x^2 + c$. Solving for y, we find $y = \pm C \exp(-x^2)$ with $C = \exp(c)$. Allowing the constant C to take both positive and negative values, we obtain the explicit form of the general solution of the given ODE: $y = C \exp(-x^2)$. It is a family of functions parametrized by the constant C. Fig. 8.11 shows three solutions from this family. Finally, we apply the initial condition to single out the unique solution to the given IVP problem. So, we seek a function of the family such that $C \exp(0) = 2$. Hence $C = 2$ and the solution to the IVP is $f(x) = 2 \exp(-x^2)$.

☼ *Check Your Understanding.* Which of the curves in Fig. 8.11 is the graph of the solution to the IVP?

8.4.2 Lab 10: Speed of a skydiver.

Problem formulation. A skydiver opens a parachute when he/she has reached a speed v_0. The total mass of the diver and parachute is m. Assume that the air resistance u (the drag force) is proportional to the square of the velocity, that is, $u(t) = C v^2(t)$ for some constant C.[8]

---

[7]Multiplication by dx is a legitimate operation since the left-hand side of equation (8.9) is a ratio.

[8]Since the motion of the diver is along a line in one direction, we will use the terms "speed" (with notation s) and "velocity" interchangeably.

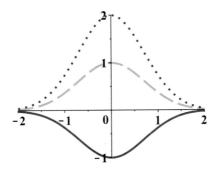

Figure 8.11. Three solutions to the separable ODE in Example 8.8

Remark 8.9. In practice, the constant C in the expression for air resistance is determined via physical considerations, such as the drag coefficient, area of the falling object, and the density of the fluid medium (in this case, air).

Do the following.

Step 1. Using Newton's Second Law, show that the motion of the skydiver can be modeled by the IVP

$$s' = \frac{C}{m}(k^2 - s^2), \quad s(0) = s_0,$$

where $k = \sqrt{mg/C}$. Find the speed as a function of time and the parameters of the problem.

Step 2. Make a function **diver_speed**$(m, C, s0)$ that takes the total mass m of a skydiver and a parachute, the coefficient C of the drag force, and the initial speed s_0, and returns the speed as a function of time. Use the general symbolic solution that you found in Step 1.

Step 3. Test your function in Step 2 for the following data (in SI units): weight $= mg = 720$, $g = 9.8$, $b = 25$, $s0 = 10$.

Answer:
$$s(t) = \frac{5.367(1 + 0.302 \exp(-3.652t))}{1 - 0.302 \exp(-3.652t)}.$$

Step 4. Calculate the limit as $t \to \infty$ of the speed function found in Step 3.[9]

Step 5. Plot the velocity and acceleration for the parameters of the test problem in Step 3 in one figure for $0 \le t \le T$ with T large enough to visually obtain approximately constant velocity.

Step 6. (Important for skydiver safety). Estimate the maximal total weight for which the landing speed is less than 6 meters per second. Assume that all parameters but the total mass are specified as in the the test problem in Step 3. Also assume that there is enough time before landing to reach (approximately) terminal speed.

8.4.3 Mini project: Projectile motion.

---

[9]This limit is called "terminal velocity" or "terminal speed".

Problem formulation. A material particle of mass m is thrown from a height H meters above the ground towards a fence of height h meters located l meters from the initial projectile location. The magnitude of the initial velocity is s m/s. Find the angle α of the horizontal at which the projectile should be thrown in order to just clear the top of the fence. Plot the trajectory of the projectile for the specified parameters given below. If there is more than one trajectory satisfying the conditions of the problem, plot all of them in one figure. Assume that the effect of air resistance is negligible. Use $g = 9.81$ m/s^2 for gravitational acceleration.

Suggested plan and directions.

Step 1. Solve the problem with specific parameters $H = 2, h = 3.5, l = 20, s = 15$ step-by-step.

(a) Use Newton's Second Law which relates the force $\mathbf{F} = [0 \ -mg]$ acting on the projectile and acceleration vector $\mathbf{a} = [a_x \ a_y]$.

$$m\mathbf{a} = \mathbf{F}$$

to set up the ODEs for the horizontal and vertical components of position vector $[x(t) \ y(t)]$ of the projectile as functions of time.[10] Find the general solutions of the ODEs.

(b) Find the integration constants in the general solutions found in Part 1 for the horizontal and vertical components of motion using
   - the initial position components $x(0) = 0, y(0) = H$
   - the initial velocity components in terms of the unknown initial direction α and the initial speed s

   Write the corresponding equations for $x(t)$, $y(t)$. Notice that the equations involve the unknown initial angle α.

(c) Use the equation for $x(t)$ to find the time t_{fence} when the projectile reaches the fence location, that is, when $x(t) = l$. The expression for t_{fence} involves the unknown parameter α.

(d) Substitute $t = t_{fence}$ into the vertical component and set the result equal to the height of the fence. Rewrite the equation as a quadratic with respect to $\tan^2(\alpha)$. The trigonometric identity $1/\cos^2(\alpha) = 1 + \tan^2(\alpha)$ would be helpful for this transformation. Solve the resulting equation and find the initial direction(s) α.

(e) Now the parametric equations $x(t)$, $y(t)$ of the required projectile trajectory(-ies) are complete. Make a figure with plot(s) of the trajectory(-ies). It is known from Galileo's time that when air resistance is negligible, a projectile trajectory is a parabola. Convert the found parametric solution to an explicit form to demonstrate this fact.

Step 2. Make a function **projectile_motion(H, l, h, s)** that takes the problem parameters and returns the angle(s) α and the parametric equations of the projectile trajectory(-ies), if the problem is solvable. Otherwise, your function should return the string "No solution". Test your function using the parameter values specified in Part 1. The angle(s) α returned by your function for these parameters should be $\alpha 1 \approx 0.992$, or $56.8°$ and $\alpha 2 \approx 0.654$ or $37.5°$.

---

[10]Recall that the components $a_x \ a_y$ of the acceleration vector are the second derivatives of the position coordinates.

Step 3. It may be the case that for some parameter values there is no solution. Set $H = 0$, $h = 3$. Find the region in the plane of the two parameters l, V for which the problem is solvable. Hint: Analyze the discriminant of the quadratic equation for $\tan(\alpha)$.

8.5 Glossary

- **Fundamental Theorem of Calculus – Part 1:** Given an integrable function f on $[a, b]$, the function

$$F(x) = \int_a^x f(t)\,dt$$

  is an **antiderivative** of f on $[a, b]$. **Part 2:** Given an antiderivative Φ of f on $[a, b]$,

$$\int_a^b f(t)\,dt = \Phi(b) - \Phi(a).$$

- **Mean value of a function over an interval** $[a, b]$:

$$f_{ave} = \frac{1}{b - a} \int_a^b f(t)\,d\,t.$$

- **Shear transformation (map) in the horizontal direction** – A transformation defined by the matrix

$$S = \begin{bmatrix} 1 & k \\ 0 & 1 \end{bmatrix},$$

  where k is a parameter called the **shear factor**. It maps the radius-vector of a point (x, y) to the radius-vector of the point $(x + ky, y)$.

- **Lorenz curve** – The curve $y = L(x)$, where the independent variable x, $0 \le x \le 1$, represents the lowest fraction of the population of a society in terms of wealth and $y(x), 0 \le y \le 1$, is the fraction of the total wealth that is owned by that fraction of the society.

- **Gini index** – A number defined as the ratio $G = A/(A + B)$, where A is the area of the region between $y = x$ and a Lorenz curve $y = L(x)$ and B is the area of the region between $y = L(x)$ and the x-axis.

9

Approximating Zeros of Functions by Iteration Methods

Many mathematical, scientific, and engineering problems boil down to solving an equation of the form $f(x) = 0$. However, such an equation can be complex to solve directly or it may not even be possible to find an explicit formula for the solution. Fortunately, in applications, it is usually not necessary to have an exact solution and a close approximation of the solution suffices. Numerical root-finding algorithms are designed to do just this. The goal of a root-finding method is to numerically approximate solutions to an equation of the form $f(x) = 0$.

The simplest of such numerical procedures is the bisection method, which was introduced in Chapter 7. This method is reliable since it is always convergent, meaning that one can guarantee that the approximation moves closer towards the real solution as the algorithm proceeds. However, it is inefficient in the sense that it may take a long time to obtain a sufficiently accurate solution. This is one motivation for exploring other methods.

A root-finding problem $f(x) = 0$ can be recast into a so called **fixed point problem** $g(x) = x$. This can be done in many ways, in particular, by adding x to both sides of the equation and denoting $f(x) + x$ as $g(x)$ (although this is not always the best trick in practice). The solution to a fixed point problem, called a **fixed point** of the function g, is found using a recursive (iterative) process of the form

$$x_{n+1} = g(x_n), \; x_0 = a, \; n = 1, 2 \ldots,$$

where a is an **initial guess**.

In computational mathematics, an iterative process, in a broad sense, is a mathematical procedure that uses an initial guess to generate a sequence of improving approximate solutions for a problem. The sequence is often written in the form $X_{k+1} = F(X_k)$. In general, X and F can be different mathematical objects. For example, X can be a symbolic vector and F a multivariate vector-function. Iteration methods are often the only choice for solving nonlinear equations. Various types of iteration methods have been developed for approximating solutions to systems of equations. These approximation methods are topics in numerical analysis courses where delicate questions of convergence, rates of convergence, and implementation issues are studied.

Our goal in this chapter is more modest. We consider only an iteration method for finding a fixed point of a scalar univariate function, as well as an application of this method called Newton's algorithm. Lab 11, one of the two labs in this chapter, is based on the mathematics used in a classic astronomy problem called the "Kepler problem." This lab includes root finding as a subproblem.

Iterative processes and their analysis are also important in studying discrete dynamical systems (DDS) that are used to model the long term behavior of various processes in science. Lab 12 in this chapter is on a heuristic investigation of the behavior of a classical DDS – the family of logistic maps.

The important issues in implementing iterative methods are iteration **convergence** and a **stopping (termination) criterion**. The former is addressed in Section 9.1 where we will state and prove a theorem that provides a sufficient condition for convergence of iteration procedure. The termination criterion could be a predefined number of iterations or sufficiently small difference between consecutive iterates.

An iteration process can be encoded as an iterative function or as a recursive function. An iterative function is one that loops to repeat some part of the code, while a recursive function calls itself during the execution. As an example, pseudocode for the recursive function FibRec(n) that takes a number $n \in N^0$ and returns the nth term of the famous Fibonacci sequence is shown below. (Recall that the Fibonacci sequence is defined by the following rule: $f_0 = 0$, $f_1 = 1$, $f_n = f_{n-1} + f_{n-2}$.)

Algorithm Finding nth term of Fibonacci sequence

1: **function** FIBREC(n)   ▷ The term f_n
2:     **if** $n < 2$ **then**
3:         **return** n ▷ Return f_0 or f_1 if condition holds
4:     **return** FibRec($n-1$)+FibRec($n-2$)

9.1 Fixed point iteration method

In this section we present and analyze the iteration method for a scalar fixed point problem $g(x) = x$. An iteration process

$$x_{n+1} = g(x_n), \ x_0 = a, \ n = 1, \ 2 \ldots \tag{9.1}$$

is called **convergent** for a given initial guess if the sequence of iterates x_n converges to a fixed point. A theorem on a sufficient condition for convergence of an iteration process is stated and proved below. The proof uses both the IVT and the MVT, which were reviewed in Chapter 7.

The next example demonstrates the "mechanics" of the iteration method using a simple root-finding problem.

Example 9.1. Convert the root-finding problem $f(x) = 0$ with $f(x) = 0.9x - 1.9x^2$ into a fixed point problem. Choose $x_0 = 0.2$ as the initial guess and calculate a sufficient number of iteration points to identify a pattern in the process dynamics.

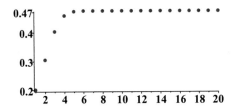

Figure 9.1. Fixed-point iterations for Example 9.1 with $x_0 = 0.2$ quickly converge to the actual root.

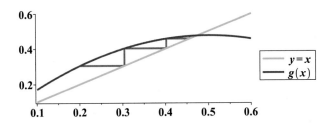

Figure 9.2. Cobweb for Example 9.1 with $x_0 = 0.2$.

Solution. Clearly, the nonzero root is $9/19 \approx 0.4736842$. We transform the equation into the fixed point problem $x = g(x)$ with $g(x) = f(x) + x = 1.9\,x(1 - x)$. Then the fixed point iteration process is stated as

$$x_{n+1} = 1.9x_n(1 - x_n), \quad x_0 = 0.2.$$

Using CAS assistance, we find a sequence of eight iteration points:

$$(0.2, 0.304, 0.40201, 0.45676, 0.47145, 0.47345, 0.47366, 0.47368, 0.47368).$$

Fig. 9.1 and this calculation show that the iterates quickly converge to the exact root. Another way to visualize the dynamics of the iterations is through what's called a **cobweb plot**, see Fig. 9.2. The value x_{n+1} in the figure is the x-value of the horizontal projection of the point $(x_n, g(x_n))$ on the line $y = x$.

One drawback to using an iteration method is that it can potentially fail. Here is an example.

Example 9.2. Consider the iteration $x_{n+1} = x_n^3 - 2$. The corresponding root-finding problem is $x^3 - x - 2 = 0$ has zero $x \approx 1.52138$. With an initial guess of $x_0 = 1.4$, we calculated the following five iterations:

$$1.4, \ 0.744, \ -1.588, \ -6.006, \ -218.628.$$

This pattern continues and shows that the iterations decrease without bound.

☼ **Check Your Understanding.** Make a cobweb for this example similar to Fig. 9.2, but now for the case of a diverging iteration process.

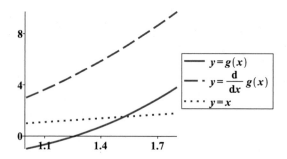

Figure 9.3. Graphical verification of the convergence condition in a neighborhood of the fixed point in Example 9.2. The convergence condition on the derivative values does not hold.

We will explain the reason of this failure after proving the following theorem, which gives sufficient conditions for the existence of the unique fixed point and convergence of the fixed point iteration process to that point.

Theorem 9.3. *Let $g : [a, b] \to [a, b]$ be a differentiable function and assume that there exists $m < 1$ such that*

$$|g'(x)| \le m \text{ for all } x \in [a, b]. \tag{9.2}$$

Then g has exactly one fixed point $p \in [a, b]$, and the sequence x_n defined by the process (9.1) *with any starting point $x_0 \in [a, b]$ converges to p.*

Proof. We split this proof into three parts.

Existence. If $g(a) = a$ or $g(b) = b$, then g automatically has a fixed point in $[a, b]$. Assume that $g(a) > a$ and $g(b) < b$. Consider the function $f(x) = g(x) - x$. Then $f(a) > 0$ and $f(b) < 0$ and we conclude by the IVT that f has a zero in (a, b). This zero is a fixed point of g.

Uniqueness. Suppose there are two fixed points, $p_1, p_2 \in [a, b]$, that is, $p_j = g(p_j)$, $j = 1, 2$. Then using the MVT and condition (9.2) yields

$$|p_1 - p_2| = |g(p_1) - g(p_2)| = |g'(c)| \, |(p_1 - p_2)| \le m |p_1 - p_2|,$$

where c is some number between p_1 and p_2. This contradiction[1] proves the uniqueness of the fixed point.

Convergence. We will use a proof technique called **mathematical induction** to prove the statement

$$|x_n - p| < m^n (b - a). \tag{9.3}$$

The technique consists of two steps:

(1) The initial or base case: Prove that the statement holds for $n = 1$.

(2) The inductive step: Assume that the statement holds for some arbitrary natural number n, and prove that the statement holds for $n + 1$.

---

[1] Remember? $m < 1$!

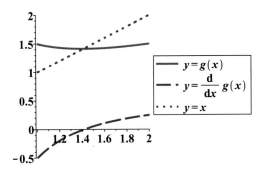

Figure 9.4. Graphical verification of the convergence condition in a neighborhood of the fixed point in Example 9.5. On the interval $[1, 2]$ the convergence condition is met.

Base case: Let $x_0 \in [a, b]$ be an initial guess. We have

$$|x_1 - p| = |g(x_0) - g(p)| = |g'(c)| \cdot |x_0 - p| \le m|x_0 - p| < m(b - a),$$

where c is some number between p and x_0.

Inductive step: Suppose the inequality (9.3) is true for a natural number n. Then

$$|x_{n+1} - p| = |g(x_n) - g(p)| = |g'(c)| \cdot |x_n - p| < m \cdot m^n(b - a) = m^{n+1}(b - a).$$

Here c is some number between x_n and p.

This completes the proof of (9.3). It follows that $x_n \to p$ as $n \to \infty$ since $m^n \to 0$. □

Remark 9.4. Using the argument in the convergence proof of Theorem 9.3, one can show that if $|g'(x)| \ge m > 1$, then the iteration process diverges. This explains why the iteration process in Example 9.2 diverges. We see from Fig. 9.4 that the derivative of the function g is strictly greater than say, two, in a neighborhood of the fixed point.

Example 9.5. Consider the iteration process $x_{k+1} = g(x_k)$ with $g(x) = \frac{1}{2}(x + \frac{2}{x})$.

(a) Plot in one figure $y = x$, $y = g(x)$, $y = g'(x)$, and use the figure to identify an interval where condition (9.2) holds.

(b) Solve the equation $x = g(x)$ exactly to find the fixed point.

(c) Run the iteration $x_{k+1} = g(x_k)$ and find six iterates of the fixed point.

Solution.

(a) See Fig 9.4.

(b) Solutions to the equation $x = g(x)$ are $\pm\sqrt{2}$.

(c) Six iterates with the initial point $x = 1.0$ are

(1.0, 1.500000000, 1.416666666, 1.414215686, 1.414213562, 1.414213562).

A high accuracy value of $\sqrt{2}$ obtained with CAS assistance is 1.4142135624.

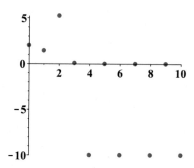

Figure 9.5. The iterative process goes into an infinite loop.

Exercise 9.6. Design a function for running an iteration process to approximate $\sqrt{5}$ and follow the plan implemented in Example 9.5. Your figure will be similar to Fig 9.4.

Example 9.7. Compare the behavior of two different fixed point iteration processes corresponding to the root-finding problem $x^4 - x - 10 = 0$.

(a) $x_{k+1} = g(x_k)$ with $g(x) = 10/(x^3 - 1)$, $x_0 = 2$.

(b) $x_{k+1} = s(x_k)$ with $s(x) = (x + 10)^{0.25}$, $x_0 = 3$.

Solution. First, let us check that each of the fixed point problems formally corresponds to approximating a root of the given root-finding problem. For part (a), we have $p = 10/(p^3 - 1) \Rightarrow p^4 - p - 10 = 0$. For part (b), we have $p = (p + 10)^{0.25}$. Raising both sides of the equation to the fourth power, we retrieve the given root-finding problem.

(a) Fig. 9.5 shows ten points obtained with CAS assistance through iterating with the function g and initial guess $x_0 = 2$. Clearly, there is no convergence.

(b) The sequence $x_{k+1} = (x_k + 10)^{0.25}$, $k = 1, \ldots, 10$, with initial guess $x_0 = 3$, quickly converges to the first real zero of the given polynomial:

  (3, 1.8988299, 1.8572743, 1.8556506, 1.8555871, 1.8555846, 1.8555845, 1.8555845).

  High accuracy approximations of the two real solutions of the given polynomial equation obtained with CAS are $x_1 = 1.8555845$, $x_2 = -1.6974719$.

Exercise 9.8. Modify the fixed point problem in Example 9.7(b) to design a fixed point iteration process that converges to the root $x = -1.6974719$. Use CAS to implement several steps of the modified iteration process. Make a figure similar to Fig. 9.1.

🔆 *Check Your Understanding.* Show algebraically or graphically that in Example 9.7(a) the condition (9.2) is violated.

Exercise 9.9. For each of the functions f below

  (i) $f(x) = x - \sin(x) - 1/2$

  (ii) $f(x) = x - \exp(-x)$

(iii) $f(x) = \exp(-x)(x^2 - 5x - 2) + 1$

consider the root-finding problems $f(x) = 0$ and do the following.

(a) Transform the equation $f(x) = 0$ into a fixed point problem of the form $x = g(x)$.[2]

(b) Plot the line $y = x$, the curve $y = g(x)$, and the function $y = g'(x)$ near the intersection point of the curve and the line in one figure.

(c) By visual inspection of the figure made in part (b), estimate the location of the fixed point and check if the condition (9.2) holds in a neighborhood of the point. If this is not the case, try a different function g for running the corresponding iteration process.

(d) Make a CAS function **my_iterates(g, x0, n)** that takes a function g, an initial guess x_0 chosen from a neighborhood of the fixed point where the condition (9.2) holds, and a natural number n. The function returns a sequence $x_{k+1} = g(x_k)$, $k = 0, \ldots, n - 1$. Test the function on an example of your choice and apply it to each of the three fixed-point problems constructed in parts (a)-(c).

9.2 Newton's method

Newton's method is a particular case of the iteration procedure that involves the derivative of the function of interest. Therefore, the method applies only to differentiable functions. The basic idea of the method is straightforward. Fig. 9.6 shows a graphic illustration of two iteration steps of Newton's method. The approximation x_{k+1} of the zero of a function f is the x-intercept of the tangent line to the graph of the function passing through the point $(x_k, f(x_k))$. Symbolically implementing this geometric idea, we arrive at the iterative formula for Newton's method:

$$x_{k+1} = x_k - \frac{f(x_k)}{f'(x_k)}. \tag{9.4}$$

Exercise 9.10. Show that the conditions

- f has a continuous second derivative in a neighborhood of its zero p

- $f'(p) \neq 0$

imply that the function $g(x) = x - f(x)/f'(x)$ that defines the iteration process (9.4) satisfies assumption (9.2) of Theorem 9.3.

Formula (9.4) can be also derived analytically using an approximation of the function f by its Taylor polynomial of degree one. In fact, suppose that there is a sequence $\{x_k\}_{k=1}^{\infty}$ that converges to the zero $x = p$ of the function f. Then

$$0 = f(p) \approx f(x_{k+1}) = f(x_k) + f'(x_k)(x_{k+1} - x_k).$$

Equating the left- and right-hand sides of this equation and solving for x_{k+1}, we obtain the iterative formula (9.4).

🔆 **Check Your Understanding.** Suppose that the iterates x_k found according to formula (9.4) converge to the point p and that f' is continuous in some neighborhood of this point. Show that the fixed point is a root of the function f.

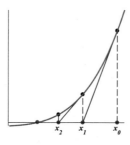

Figure 9.6. The geometric idea of the Newton's method.

Next we apply Newton's method to a simple example that was used earlier in demonstrating the bisection method.

Example 9.11. Use Newton's method to approximate the root of the function $f(x) = x - \cos(x)$ on the interval $[0, 1]$.

Solution. Using CAS, we implement three steps of the iteration process

$$x_{k+1} = x_k - \frac{x_k - \cos(x_k)}{1 + \sin(x_k)}$$

with initial guess $x_0 = 0.2$:

$$x_1 = 0.7394364979, \quad x_2 = 0.7390851605, \quad x_3 = 0.7390851332.$$

The "exact" value of the root is 0.7390851332. Observe that Newton's method quickly (in just three iterations!) converges to the "exact" solution, which is a striking difference compared to the slow convergence seen using the bisection method.

Newton's method is fast but it has disadvantages too:

- iterates may diverge

- the method uses the derivative of the function of interest, so the function must be at least differentiable

- there is no practical and rigorous error bound

Fig. 9.7 shows how a bad choice of the starting point results in the failure of Newton's method. The iterate x_1 jumps away beyond the domain of the given function. Once one finds a root isolation interval, the rule of thumb for choosing an initial guess x_0 for Newton's method is the condition $f(x_0)f''(x_0) > 0$.

🔆 **Check Your Understanding.** By visual inspection, determine the sign of the product $f(x_0)f''(x_0)$ for the function depicted in Fig. 9.7.

📖 Newton's method generalizes to complex functions and systems of equations.

---

[2]Be aware that this transformation can be done in different ways.

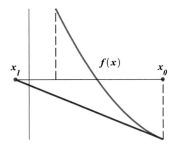

Figure 9.7. Failure of Newton's method. The first iterate is beyond the function domain.

9.3 Lab 11: Kepler's Equation and deriving Kepler's Second Law

Kepler's three laws of planetary motion describe the motion of planets in the solar system:

- **First Law (law of ellipses)**: The orbit of every planet is an ellipse with the Sun at one of the two foci.

- **Second Law (law of equal areas)**: A line joining a planet and the Sun sweeps out equal areas during equal intervals of time.

- **Third Law (law of harmonies)**: The square of the orbital period of a planet is proportional to the cube of the semi-major axis of its orbit.

Kepler's efforts to explain the underlying reasons for such motions are no longer accepted. Nonetheless, the actual laws themselves are still considered an accurate approximation of the motion of any planet and any satellite. Kepler's laws represent an idealized model; it is a particular case of the so called **two-body problem**. In reality, other celestial bodies cause perturbations of Kepler's model.

📖 Isaac Newton developed a more general set of principles which applies not only to the heavens but also to the earth in a uniform way. In his book **Philosophiae Naturalis Principia Mathematica**, Newton stated three laws of motion and the law of universal gravitation. From these laws Newton was able to derive all of Kepler's laws of planetary motion. The simplicity and extremely broad applicability of Newton's approach forever changed both astronomy and mathematics.

Terminology and notation.

- **Eccentricity e of the ellipse**: $e = c/a = \sqrt{1 - (b/a)^2}$. Note that $b = a\sqrt{1 - e^2}$.

- **Mean anomaly**: $m(t) = 2\pi t/T$.

- **Eccentric anomaly** $E(t)$: An angular parameter that defines the position of a body moving along an elliptic Kepler orbit (see details below).

- **Kepler's equation**: $m(t) + e \cdot \sin\big(E(t)\big) = E(t)$.

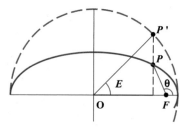

Figure 9.8. Geometric meaning of the eccentric anomaly E.

The term **eccentric anomaly** might look new to mathematics students, but they use it (without knowing its name) every time they encounter the standard parametric equation of an ellipse $x = a\cos(t), y = b\sin(t)$. The conventional mathematical notation t here is what is called in astronomy the eccentric anomaly and is typically denoted by E. Fig. 9.8 explains the geometric meaning of this term. In the figure, the red solid curve is a fragment of an elliptic planet orbit. The blue dashed curve is an arc of the circle of radius a. The point P is the position of the planet, and the point F the right focus of the planet orbit.

Problem formulation. Suppose a planet orbits around the Sun along the elliptic orbit $x^2/a^2 + y^2/b^2 = 1$ with orbital period T. Assume that the Sun is in the right focus $F(c, 0)$ of the orbit. Choose the right focus of the orbit as the pole of the polar coordinate system, and measure the polar angle θ from the polar axis that goes along the x-axis.

Do the following.

Part 1. Derive parametric equations $x = x(E)$, $y = y(E)$ for the orbit with the eccentric anomaly E as the parameter.

Directions: Use geometric considerations to express the x-coordinate of the position P of the planet in Fig. 9.8 in terms of E. Find an expression for the y-coordinate using the standard equation of an ellipse.

Part 2. Assume that at $t = 0$ the planet is at the point $(a, 0)$, the closest position to the Sun.[3] Make a CAS function **positions**$(\mathbf{a}, \mathbf{e}, \mathbf{T}, \mathbf{n})$ that takes parameters a, e, T, n and returns

- the list *Evals* of the values of the eccentric anomaly $E_k = E(t_k)$, $t_k = T \cdot k/n$, $k = 0, \ldots, n-1$.
- the list P of corresponding positions of the planet
$$P = [(a, 0), (x(E_1), y(E_1)), (x(E_2), y(E_2)), \ldots, (x(E_{n-1}), y(E_{n-1}))].$$

Directions: Kepler's equation does not have a closed-form solution for $E(t)$. Use a built-in root-finding CAS function to find approximations $E(t_k)$ of the values in the list *Evals*. Test your function with parameters $e = 0.7$, $a = 2$, $n = 5$, $T = 5$.

Answer to the test problem:

$$Evals = [0, 1.915, 2.7681, 3.515, 4.368]$$

$$P = [(2, 0), (-0.676, 1.344), (-1.8620, 0.521), (-1.862, -.521), (-0.676, -1.344)]$$

---

[3] In astronomy the point is called the **perihelion.**

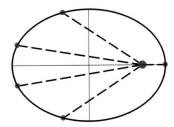

Figure 9.9. Elliptic orbit and positions of the planet with orbital period $T = 5$ at times $t = k, k = 0, \ldots, 5$.

Part 3. Use the function made in Part 2 with parameters $e = 0.7$, $a = 2$, $n = 7$, $T = 7$ to plot the elliptic orbit, the Sun located at the right focus of the orbit, and the positions of the planet at times $t = t_k, k = 0, \ldots, 6$, in one figure. The figure will be similar to Fig. 9.9.

Part 4. Use the same parameters as in Part 3 to demonstrate Kepler's Second Law by calculating areas A_k, $k = 1, 2, \ldots, 7$ swept by the polar radius of the planet during each of the seven time intervals.

Directions: It is known that the polar equation of an ellipse with polar center at the right focus is $r(\theta) = a(1 - e^2)/(1 + e\cos(\theta))$. Use Fig 9.8 and geometric considerations to find the limits of integration θ_k in terms of the eccentric anomaly values E_k. **Hint**. Notice that the polar angle in Fig 9.8 equals

$$\theta = \pi - \arctan\left(\frac{y(E)}{c - x(E)}\right) \tag{9.5}$$

This formula defines θ as a function of E, $\theta = g(E)$. Thus, the limits needed for calculating areas A_k swept during equal intervals of time are $0, \theta_k = g(E_k), 2\pi$. Each A_k can then be calculated as the area of the polar region bounded by the graph of $r = r(\theta)$ between the radius-vectors $\theta = \theta_k$ and $\theta = \theta_{k+1}$:

$$A_k = \frac{1}{2} \int_{\theta_k}^{\theta_{k+1}} r^2(\theta) \, d\theta.$$

Use CAS assistance for calculating these integrals.

💡 **Check Your Understanding.** Justify formula (9.5) for all E-values.

Part 5. Use the chain rule to show that

$$\frac{dA}{dt} = C, \tag{9.6}$$

for some constant C and explain why the Kepler's Second Law follows from this equation.

Suggestion: To derive the equation (9.6), notice that the area swept by the polar radius from $t = 0$ to time t is

$$A(t) = \frac{1}{2} \int_0^{\theta(E(t))} r^2(s) \, ds,$$

where the eccentric anomaly E is implicitly defined as a function of t by Kepler's equation. Thus $A = A(\theta(E(t)))$, and therefore, by the chain rule,

$$\frac{dA}{dt} = \frac{dA}{d\theta}\frac{d\theta}{dE}\frac{dE}{dt}.$$

Use Kepler's equation and implicit differentiation to find dE/dt. All derivatives should be functions of the eccentric anomaly.

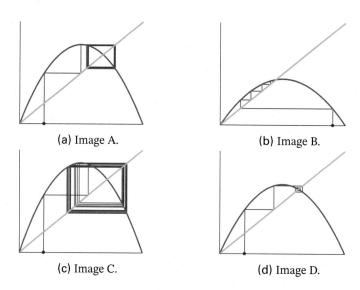

(a) Image A.   (b) Image B.

(c) Image C.   (d) Image D.

Figure 9.10. Four logistic maps.

9.4 Lab 12: Exploration of the logistic maps

The family of logistic maps is a popular example used to show how complex behaviour can arise from very simple nonlinear polynomial dynamical equations. The family is defined as $f_r(x) = rx(1-x)$, $0 \le x \le 1$, $0 < r \le 4$.

Problem formulation

Part 1. Make a function *logistic_cobweb*(r, xini, n) that takes a value of the parameter r, an initial x-value $x_0 \in (0,1)$, and the number of iterations n, and returns a plot of iteration points $(x_k, f_r(x_k))$, $k = 1, \ldots, n$ with $x_k = f_r(x_{k-1})$.

Part 2. Apply your function to logistic maps with the following r-values:

$$r_1 = 0.95,\ r_2 = 1.9,\ r_3 = 2.9,\ r_4 = 3.1,\ r_5 = 3.5. \tag{9.7}$$

For each of the given values of the parameter r, start with some initial value x_0 and implement iterations of the logistic map with this parameter until you develop some understanding of the behavior of this particular logistic map. Experiment with initial values and the number of iterations to produce an informative picture.

Part 3. Fig. 9.10 shows cobwebs A, B, C, D of four logistic maps with parameter values from the list (9.7). The blue curve is the graph of the logistic map $f_r(x)$, the green line is the graph of the line $y = x$, and the red points have the coordinates $(x_0, f_r(x_0))$. Based

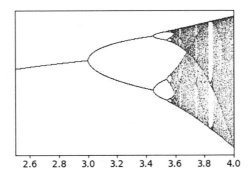

2.6   2.8   3.0   3.2   3.4   3.6   3.8   4.0

Figure 9.11. Bifurcation diagram for the family of logistic maps.

on your results in Part 2, match each of the figures with one of the parameter values in (9.7).

📖 There is a way to graphically summarize long term behavior of the logistic family for the range of parameter values $0 < r \leq 4$ using a so called **bifurcation diagram** (see Fig. 9.11). Typically, the diagram is constructed as follows. A large number of equidistant values of the parameter r in the interval $(0, 4]$ are generated, and a large number of iterations are implemented for each of these values. To only capture the long term (asymptotic) behavior, a significant number of iterates at the beginning of the iterative process are discarded. If iterations of a particular function f_r converge to a limit $l(r)$, the point $(r, l(r))$ is added to the diagram. A function with two-periodic asymptotic behaviors adds two points to the bifurcation diagram, and so on. For $r > 3.54409$ (approximately) the asymptotic behavior becomes more complicated.

9.5 Glossary

- **Fixed point problem** – Finding a solution to an equation of the form $g(x) = x$. The solution is called **fixed point**.

- **Fixed point iteration** – The process of finding a solution to a fixed point problem using an equation of the form $x_{k+1} = g(x_k), k = 0, 1, \ldots$, with an initial value x_0 called the **initial guess**. Under certain conditions, the process produces successively better approximations to the solution of the fixed point problem $g(x) = x$.

- **Newton's method** – A root-finding algorithm that uses a certain fixed point iteration to find successively better approximations to a root-finding problem for a differentiable function.

- **Bifurcation diagram** – A display summarizing long term (asymptotic) behavior of a family of maps depending on a parameter. This **bifurcation parameter** is used in the diagram as an independent variable in conventional graphs of functions, but for systems of nonlinear maps the diagram can show multiple outputs, as in Fig. 9.11.

10

Polynomial Approximations

In many applications it is necessary to perform mathematical operations on a function that is defined by a table of values obtained via some measurements or a function that is too complex to efficiently work with. Examples of such operations include function evaluation, function integration, prediction of function values for new inputs when the function is defined by a table, or solving a differential equation for a function that is implicitly defined by this equation.

A set of points for a function of interest f can be used to construct a model g of this function. It is important to ensure that a mathematical operation we want to apply to f can be performed on the function g and provide useful information about f. In this case, we call the function g an **approximation** of f. If, in addition, g must have the same values as f at certain points, we call the function g an **interpolant** of f and the process of its construction **interpolation**.

Polynomials and trigonometric functions are commonly used as building blocks for constructing function approximations and interpolants. A polynomial approximation can be built as a finite linear combination of certain polynomials $T_k(x)$ called **basis functions**:

$$P_n(x) = a_0 + a_1 T_1(x) + a_2 T_2(x) + \cdots + a_n T_n(x). \tag{10.1}$$

The basis polynomials are carefully chosen for making the approximation procedure effective and efficient for the problem at hand. The approximation (10.1) is completely defined by a finite list of coefficients $[a_k, k = 0, 1, \ldots, n]$. This provides an opportunity to use computer assistance in both building the approximations and implementing the needed mathematical operations on P_n instead of the function of interest. When the basis functions are powers of the independent variable, the equation (10.1) becomes just the standard form of a polynomial in powers of x. There are other choices for systems of basis functions that lead to even non-polynomial approximations. The so called trigonometric system is one of such systems. It will be introduced in the next chapter.[1]

---

[1] Some topics and exercises in this chapter and the next one are prompted by presentation of the approximation theory in [6].

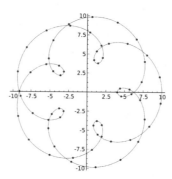

Figure 10.1. An interpolant of a parametric curve of mechanical descent.

How "good" can a polynomial approximation of a function be? The famous Weierstrass theorem gives a surprising answer: if a function f is continuous on some closed and bounded interval, then for any band surrounding the function graph, no matter how narrow, there exists a polynomial with the graph completely inside the band. Although very impressive, the theorem is a pure existence statement, and its proof does not provide a practical way for constructing a polynomial approximation for a given band.

Piecewise polynomial interpolants called splines are often preferred in interpolation problems. They are more flexible in interpolating complicated curves and avoid the unwanted Runge phenomenon (which we describe later in this chapter). Parametric splines are routinely used in applications such as computer aided engineering design and computer graphics, but this spline technique is beyond the scope of this chapter. A nice picture in Fig. 10.1 shows a cubic spline with a large number of "pieces" defined by appropriate cubic polynomials. The spline interpolates a rather exotic curve defined parametrically.

When the number of data points is large, a polynomial interpolation could produce a high-degree polynomial with many twists and turns. If a model for a large data set follows the data too dutifully, it may not work well for new, unseen data. In data science, this phenomenon is called **overfitting**. Modeling based on large sets of data typically use the so called **Least Squares (LS)** technique. For modeling a relationship between a **predictor variable** x and a **response variable** y, the LS method uses a list of data points $[(x_k, y_k), k = 1, \ldots, n]$ and a certain functional form of an approximating function chosen by the user. The functional form involves some unspecified parameters. Once the parameters are found using the LS method, the model can then be used to predict values of the response variable for new, unseen values of the predictor variable.

To find the model based on the given data, the LS method employs minimization algorithms on the sum of squares of so called **residuals** – differences between y-values given in the data and predicted by the model. One of the most important application of the LS technique is for data fitting in statistics.

In this chapter, we introduce and demonstrate some techniques of polynomial approximation and interpolation. After a brief review of a few basic terms, facts, and techniques of approximation theory needed for constructing the interpolants and approximations, these constructions will be put to work in exercises and two mini projects with CAS assistance. The LS technique will be applied in two labs at the end of chapter to construct models of the relationship between certain real-life quantities.

At a deeper theoretical level approximation and interpolation techniques are studied in numerical analysis courses.

10.1 Taylor polynomials

In this section we will review Taylor polynomial approximations for simple elementary functions and analyze how good the approximations are.

10.1.1 Review: Definition and accuracy of Taylor polynomials. We used the polynomial

$$P_1(x) = f(c) + f'(c)(x - c)$$

in Chapter 7 as a linear approximation of a function f in a neighborhood of $x = c$. This polynomial is also known as the **first-order Taylor polynomial**. Better approximations can be found using higher-order polynomials called n^{th} Taylor polynomials. Here is the general definition.

Definition 10.1. If a function f has derivatives up to order $n + 1$ in a neighborhood of the point $x = c$, the nth **Taylor polynomial centered at** c of the function f is

$$P_n(x) = f(c) + f'(c)(x - c) + \frac{f''(c)}{2!}(x - c) \cdots + \frac{f^{(n)}}{n!}(c)(x - c)^n.$$

The difference $R_n(x) = f(x) - P_n(x)$ is the **remainder term** of this Taylor polynomial.

Some important constants, such as π and e can be approximated by an appropriate Taylor polynomial. For instance, any Taylor polynomial for the function $\exp(x)$ centered at $x = 0$ with input value $x = 1$ gives an approximation of the number e.

Using Taylor polynomials for approximating functions is justified by **Taylor's theorem**. The theorem claims that the remainder has the form

$$R_n(x) = \frac{f^{(n+1)}(\xi)}{(n+1)!}(x - c)^{n+1}, \tag{10.2}$$

where ξ is some number between x and c. This is the **Lagrange form of the remainder**.

Suppose all derivatives of a function f are bounded in absolute value by a constant M on some interval I and $c \in I$. Then for any $x \in I$ the approximation error of f by its Taylor polynomial P_n can be estimated as

$$error(n) \equiv |R_n(x)| \leq \frac{M}{(n+1)!}|x - c|^{n+1}. \tag{10.3}$$

The expression on the right-hand side of this equation is bounded above by $\frac{M}{(n+1)!}d^{n+1}$, where $d = |x - c|$. It is known that the series $\sum_{k=0}^{\infty} d^k/k!$ converges for any $d \in \mathbb{R}$. By the necessary condition of convergence for a numerical series, $d^k/k! \to 0$ as $k \to \infty$. This implies that $error(n) \to 0$ as $n \to \infty$. That is, the error converges to zero as the degree n of the Taylor polynomial approximation increases. Notice that equation (10.3) also shows that the closer x is to the center c, the more accurate the approximation is.

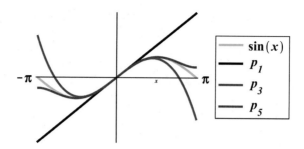

Figure 10.2. Sine function and its linear, cubic, and quintic Taylor polynomials centered at $x = 0$.

Example 10.2. Plot the linear, cubic, and quintic Taylor polynomials centered at $x = 0$ for the sine function.

Solution. Fig. 10.2 shows the three Taylor polynomials centered at the origin for the sine function. When the degree of the polynomial grows, it approximates the sine function better on larger and larger intervals.

Example 10.3. Approximate the sine function by the quintic Taylor polynomial P_5 centered at zero on the interval $I = [-\pi/2, \pi/2]$.

(a) Estimate the error using formula (10.3).

(b) Calculate the exact error at $x = \pi/2$ by finding the difference $1 - P_5(\pi/2)$.

Solution.

(a) According to (10.3), the theoretical error estimate is

$$error(5) \leq \frac{\pi^6}{2^6 \cdot 6!} \approx 0.0209, \text{ or about 2\%.}$$

(b) The quintic Taylor polynomial at $x = \pi/2$ evaluates to approximately 1.0045. Thus, the actual error is about 0.5%, which is much smaller than the theoretical error.

Remark 10.4. The theoretical error estimate (10.3) is valid for a broad class of functions, and it is typical for it to be an overestimate for a concrete function from this broad class.

The error estimate (10.3) can be used to decide how many terms in the Taylor approximation will ensure that a specific accuracy requirement is met. This is shown in the next example.

Example 10.5. Consider the function $f(x) = \exp(x)$. Find the smallest value of N such that

$$|f(x) - P_N(x)| \leq 0.01 \text{ for all } x \in [-2, 2].$$

Solution. Recall that the nth Taylor polynomial centered at $c = 0$ of the exponential function is $P_n(x) = \sum_{k=0}^{n} x^k/k!$. The estimate (10.3) with $M = \exp(2)$ (**why this**

value?) implies the inequality

$$error(n) \leq \frac{M}{(n+1)!} 2^{n+1}.$$

Computing the right-hand side of this inequality for several values of n, we find that

$$error(7) \leq \exp(2) \cdot 2^8/8! \approx 0.0104; \quad error(8) \leq \exp(2) \cdot 2^9/9! \approx 0.0021 < 0.01.$$

Answer: $N = 8$.

10.1.2 Mini project: Using Taylor polynomials for approximating a power function.

Intuition. Suppose we want to approximate $\sqrt{a}$ for some positive number a. Assume that $0 < a < 2$ and rewrite $\sqrt{a}$ as $\sqrt{1+b}$, $b = a - 1$. Thus, $\sqrt{a}$ is the value of the function $g(x) = \sqrt{1+x}$ with $|x| = |b| < 1$. Recall that the approximation of the function $g(x) = \sqrt{1+x}$ by a Taylor polynomial P_n centered at the origin can be made as accurate as you wish by increasing the degree n **provided** $|x| < 1$.

Now let a be any positive number. Here is a simple algebraic trick for reducing this general case to the previous one. You can always find an exact square that bounds a from above. For instance, if $a = 5$ then $a < 3^2$ and

$$\sqrt{5} = \sqrt{3^2 + 5 - 3^2} = 3\sqrt{1-b} \text{ with } b = (3^2 - 5)/3^2 < 1.$$

To use this trick for large numbers in a systematic way that can be encoded, one needs to find a pattern in choosing such an exact square. Powers of 10 will work. For example, any 1- or 2-digit input a is less than 100. Thus, we can write

$$\sqrt{a} = 10\sqrt{a/100} = \sqrt{1-b}, \quad b = (100 - a)/100.$$

The same trick works for larger inputs of the square root function.

Problem formulation.

Step 1. Construct a Taylor polynomial $P_4(x)$ for the function $g(x) = \sqrt{1+x}$ centered at the origin. Plot the function and the polynomial on the interval $[-1, 1]$ in the same figure.

Step 2. Use the polynomial $P_4(x)$ to estimate $\sqrt{6.2}$. Then find the "exact" value using an appropriate CAS command and calculate the relative error of this approximation.

Step 3. Make a CAS function *root2*$(\mathbf{a}, \mathbf{n})$ that takes a positive real number a and an integer n and returns the approximation of $\sqrt{a}$ using an appropriate value of the Taylor polynomial $P_n(x)$ of the function g. Test the function on an example of your choice. Make a table with several values of n and corresponding **relative errors** for your example. Comment on the results of the calculations.

10.2 Interpolating polynomials in the Lagrange form

Taylor polynomials are widely used in both mathematics and applications for approximating functions *locally*. However, it is often necessary that the interpolant is accurate on a specific interval. In this section we introduce a new kind of interpolant that achieves this called the **Lagrange polynomial**.

Let $[(x_k, y_k) : x_l \neq x_m \text{ for } l \neq m]_{k=0}^n$ be a list of data points.[2] The values x_k and y_k are called **nodes** and **nodal values**, respectively. The Lagrange polynomial interpolant is the unique polynomial of degree n defined by this data as a linear combination of **Lagrange basis polynomials** $l_k(x)$, $k = 0, \ldots, n$. Each basis polynomial l_k is the polynomial of degree n such that $l_k(x_k) = 1$ and $l_m(x_k) = 0$ for $m \neq k$. The coefficient of l_k in the Lagrange interpolant is the nodal value y_k. Thus, the Lagrange polynomial interpolant is defined as

$$L_n(x) = \sum_{k=0}^n y_k l_k(x). \tag{10.4}$$

$\boxed{\cdot\!\!\cdot}$ **Check Your Understanding.** Use the definition of the basis functions and equation (10.4) to show that $L_n(x)$ agrees with f at the nodes, that is,

$$L_n(x_k) = y_k, \ k = 0, \ldots, n$$

$\boxed{\text{📖}}$ The rigorous definition of the **basis of a finite dimensional vector space** is typically introduced in linear algebra courses. A simple example is the basis $\mathbf{i} = [1\,0]$, $\mathbf{j} = [0\,1]$ for the space of plane vectors. These two vectors form a basis in the sense that any vector on the plane furnished with a rectangular coordinate system can be uniquely represented as a linear combination of them. Polynomials of degree less or equal $n \in \mathbb{N}$ form a finite dimensional space, and a basis in this space is defined by the same property. For instance, the functions $\{1, x, x^2\}$ form a basis in the space of all polynomials of degree $n \leq 2$.

Exercise 10.6. Prove that given a list of points $[(x_k, f(x_k))]_{k=0}^n$ on the graph of a function f, the Lagrange interpolant L_n is the **unique** polynomial of degree n that agrees with the function at the nodes $x = x_k$. **Hint:** Recall how many roots a polynomial of degree n has. Then the claim can be proved using a very short contradiction argument.

The next exercise involves constructing Lagrange basis polynomials of degree n. Parts (a) through (c) will be helpful for developing some intuition about this construction.

Exercise 10.7 (Constructing Lagrange basis polynomials).

(a) Consider the nodes x_0, x_1. Show that the functions

$$l_0(x) = \frac{x - x_1}{x_0 - x_1}, \quad l_1(x) = \frac{x - x_0}{x_1 - x_0},$$

form the Lagrange basis for interpolants of degree one on the interval $[x_0, x_1]$.

(b) Consider the nodes x_k, $k = 0, 1, 2$. Show that the functions

$$l_0(x) = \frac{(x - x_1)(x - x_2)}{(x_0 - x_1)(x_0 - x_2)}$$

$$l_1(x) = \frac{(x - x_0)(x - x_2)}{(x_1 - x_0)(x_1 - x_2)}$$

are the Lagrange basis polynomials on $[x_0, x_2]$. Then construct the missing basis polynomial l_2.

---

[2]In this section and the next we assume that the values x_k form an increasing sequence.

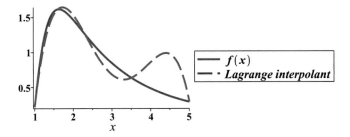

Figure 10.3. Lagrange interpolant for the function f defined in Step 4 of the mini project "Constructing Lagrange polynomial".

(c) Show that the polynomial

$$l_0 = \frac{(x - x_1)(x - x_2)(x - x_3)}{(x_0 - x_1)(x_0 - x_2)(x_0 - x_3)}$$

is the Lagrange basis polynomial $l_0(x)$ for the cubic Lagrange interpolant. Analyze the structure of this basis polynomial and construct the Lagrange cubic basis polynomials $l_j(x)$, $j = 1, 2, 3$.

(d) Complete the formula for the Lagrange basis polynomials l_k, $k = 0, \ldots, n$ of degree n:

$$l_k(x) = \prod_{j=0, j \neq k}^{n} \frac{?}{x_k - x_j}.$$

Most CASs have commands for constructing the Lagrange interpolants. However, our purpose in this text is to better understand techniques for various simple mathematical constructions and implement some of these constructions "from scratch" with computer assistance. This is what will be done in the next mini project.

10.2.1 Mini project: Constructing Lagrange polynomial.

Problem formulation. Make a CAS function *my_lagrange*(**fcn, a, b, n**) that takes a function fcn, endpoints of an interval $[a, b]$ in the function domain, and a natural number n, and constructs the Lagrange polynomial interpolant $L_n(x)$ using $n + 1$ equally spaced nodes $a < x_1 < \ldots < x_{n-1} < b$. The function should return a figure with plots of fcn and L_n on the interval $[a, b]$.

Suggested plan.

Step 1. Make a help function *get_data*(**fcn, a, b, n**) that takes a function f, the endpoints a, b of an interval, and a natural number n and returns the list of $n + 1$ data points $[(x_j, fcn(x_j)) | x_j = a + k\Delta x, \ \Delta x = (b - a)/n]_{k=0}^{n}$.

Step 2. Make a help function *lagrange_basis*(**xdata, k**) that takes x-values of $n + 1$ data points and an integer k, $0 \leq k \leq n$, and returns the basis Lagrange polynomial $l_k(x)$.

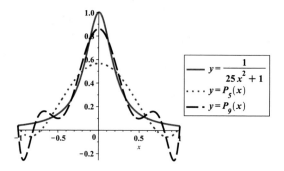

Figure 10.4. Runge phenomenon: Equally spaced nodes lead to growing interpolation error.

Step 3. Use the help functions made in Step 1 and Step 2 to make the function **my_lagrange**. Test your function for $f(x) = x/(1+x) - \cos(5/x)$ with various numbers of nodes on the interval $[1, 5]$. Your figure will be similar to Fig. 10.3.

Remark 10.8. Lagrange interpolation with equally spaced nodes can sometimes lead to growing deviations of interpolants from the function of interest as the degree of interpolants increases. Fig. 10.4 shows a classic example of this situation, called Runge's phenomenon, for the function $f(x) = 1/(25x^2 + 1)$ on the interval $I = [-1, 1]$. The 5th degree Lagrange polynomial interpolant (blue curve) is constructed using six equally spaced interpolation points on the interval I. The 9th degree Lagrange interpolant (black curve) is then constructed using ten equally spaced points on I. We see a growing interpolation error for x-values near the end points of the interval when the order of Lagrange interpolants with equidistant nodes increases. It is known from numerical analysis that one of the reasons for this unwanted effect is a quick growth of higher order derivatives of the function. There are techniques that judiciously use some freedom in choosing locations of the nodes for reducing the interpolation error.

📖 Changing the nodal points requires recalculation of the entire Lagrange interpolant, so it is often better to use *Newton's form* of polynomial interpolation, which is free from this drawback.

In the next section we will introduce a technique that leads to better interpolants than Lagrange polynomials.

10.3 Piecewise polynomial interpolation: Splines

To obtain more accurate polynomial approximations of complicated curves, an interval of interest is divided into a number of subintervals, and piecewise polynomial functions called **splines** are constructed. Splines have some advantages compared with polynomial interpolants in "one piece". In particular, they provide more flexibility to a user for designing and controlling the shapes of curves. In this section we will only work with linear and cubic splines. The input for building a spline is a set of data points, just as for Lagrange interpolants. In spline interpolation each data point is called **knot**. The

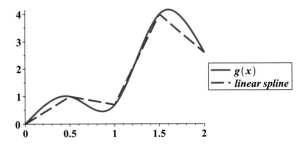

Figure 10.5. Piecewise linear interpolant of the function $g(x)$.

highest degree polynomial used in a spline is called the **spline degree**. Splines of degree three, called **cubic splines**, are most common. The simplest continuous splines have degree one. They are piecewise linear interpolants.

In Exercise 10.7(a) the Lagrange basis for linear interpolants on the interval $[x_0, x_1]$ was introduced. To use it for a piecewise linear interpolation, it is more suggestive to change the notation for the interval to $[x_{left}, x_{right}]$ and the notation for the basis functions to

$$l_{left} = \frac{x_{right} - x}{x_{right} - x_{left}}, \quad l_{right} = \frac{x - x_{left}}{x_{right} - x_{left}}.$$

Given two data points (x_{left}, f_{left}), (x_{right}, f_{right}) for a function f, the linear combination

$$f_{left} l_{left}(x) + f_{right} l_{right}(x)$$

gives you the first degree interpolant of the function on the interval $[x_{left}, x_{right}]$. This construction can be used repeatedly on each subinterval when one is building a spline of degree one.

Exercise 10.9.

(a) Manually construct the piecewise linear interpolant using the list of knots

$$[(0.0, 0.0), (1.0, 2.0), (2, 0.7), (3.0, 4.0)]$$

and make a plot of this linear spline.

(b) Make a CAS function **spline1(L)** that takes a list L of n data points and returns the linear spline through the points in the list. Test the function on the spline constructed in part (a).

(c) Use your function to construct and plot the linear spline through the points

$$[(0.0, 0.0), (0.5, 1.0), (1.0, 0.7), (1.5, 4.0), (2.0, 2.6)]$$

and plot the spline and the given points in one figure. Your figure will be similar to Fig. 10.5, but without the curve $g(x)$.

Splines of degree three are defined using data points and some conditions on the derivatives.

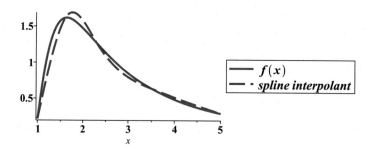

Figure 10.6. This cubic spline with six knots fits the curve in Exercise 10.12 like a glove.

Example 10.10 (About linear systems of equations for constructing cubic splines). Consider a cubic spline for the knots $[(0,0),(1,1),(2,4),(3,3)]$:

$$g(x) = \begin{cases} 0.8\,x^3 + 0.2\,x & \text{for } 0 \le x \le 1 \\ -2\,x^3 + 8.4\,x^2 - 8.2\,x + 2.8 & \text{for } 1 < x \le 2 \\ 1.2\,x^3 - 10.8\,x^2 + 30.2\,x - 55.8 & \text{for } 2 < x \le 3 \end{cases}$$

For the three cubic polynomials that define the spline we need twelve equations to find 12 coefficients (four coefficients for each of the three cubic polynomial pieces). Assigning endpoint values to each of the three polynomials gives **six conditions**. Equating the first and second derivative values of the neighboring polynomials at the inner knots gives **four conditions**. **Two** additional conditions are needed to get the required number of twelve equations for the twelve coefficients of the cubic spline. The so called **natural end conditions** are often used by default in most CASs. They assign values to the second derivative at the endpoints of the total interval.

🔅 *Check Your Understanding.* What are the values of the second derivative assigned at the end points of the total interval $[0, 3]$ to the spline g in Example 10.10?

Exercise 10.11. Use an appropriate CAS command to construct the spline in Example 10.10. Verify that at the internal knots the cubic pieces of the spline have

- the same values

- the same first derivative values (slopes)

- the same second derivative values

Plot the data points and the cubic spline in the same figure.

Any cubic spline has the properties listed in Exercise 10.11. In formal mathematical language, **a cubic spline is a twice-differentiable function**.

Exercise 10.12. Consider the function $f(x) = x/(x+1) - \cos(5/x)$ on the interval $[1, 5]$. Make a data set with six data points of your choice and use an appropriate CAS command to construct the cubic spline through these points. Plot the function and the spline interpolant in one figure. Experiment with the knot locations to get a figure similar to Fig. 10.6. (This is a spline interpolant to the same function as in Fig. 10.3, but now the interpolant fits the function much better.)

10.4 Approximating large data sets: Regression

A real data set can be too large to use polynomial or piecewise polynomial interpolation. In addition, there could be points in the set with the same x-value but different y-values. The Least Squares (LS) is a general class of approximation methods for such situations.

Suppose you have a list of observed or measured pairs of values $[(x_k, y_k)]_{k=1}^n$ of two quantities, x, the predictor variable, and y, the response variable. Let $\mathcal{F} = \{f(x, \alpha), \alpha = [\alpha_1, \alpha_2, \ldots, \alpha_n]\}$ be a certain class of single variable functions depending on one or more parameters. The goal is to find the parameters $\alpha^* = [\alpha_1^*, \alpha_2^*, \ldots, \alpha_n^*]$ such that the function $f^* = f(x, \alpha^*)$ is the best fit in the class $\mathcal{F}$ for the given data. The measure of the error between the values $Y_p = [f(x_1, \alpha), f(x_2, \alpha), \ldots, f(x_n, \alpha)]$ predicted by a function $f \in \mathcal{F}$ and the observed values $Y = [y_1, y_2, \ldots, y_n]$ is:

$$error = \sqrt{\sum_{k=1}^n \left(f(x_k, \alpha) - y_k \right)^2}. \tag{10.5}$$

Definition 10.13. The differences $y_k - f(x_k)$ between the observed values and the estimated values of the quantity of interest are called **residuals**.

Definition 10.14. The **LS best fit** from a class of functions $\mathcal{F}$ for a given list of data $[(x_k, y_k)]_{k=1}^n$ is a function $f^* \in \mathcal{F}$ that minimizes the error (10.5), or, equivalently, the radicand

$$\sum_{k=1}^n \left(f(x_k, \alpha) - y_k \right)^2. \tag{10.6}$$

This definition justifies the name "Least Squares."

10.4.1 Solving the linear LS minimization problem.
Consider the two-parameter set of linear functions $\mathcal{F} = \{f(x, a, b) = a\,x + b \mid a, b \in \mathbb{R}\}$. For functions in $\mathcal{F}$ the expression (10.6) takes the form

$$g(a, b) = \sum_{k=1}^n (a\,x_k + b - y_k)^2. \tag{10.7}$$

Then the linear LS best fit is $f^*(x) = a^* x + b^*$. The graph of this function is called the **line of best fit** or **regression line**.

It is common practice to make a **scatterplot** before choosing this class $\mathcal{F}$ of "candidates" for modeling the data. A scatterplot showing a reasonable linear association between the variables is typically considered as some justification for such a choice.

Derivation of the system for finding the optimal parameters a^*, b^* (optional).
Calculus comes into play to provide tools for finding the optimal parameters a^*, b^*. The function g depends on two variables. To derive the formulas for the solution of the minimization problem

$$\min g(a, b), \ a, b \in \mathbb{R},$$

we will need two facts from multivariable calculus.

Fact 1: Partial derivatives of a function of two variables. Let $f(x, y)$ be a function of two variables defined in some disk centered at (x_0, y_0). Partial derivatives f_x, f_y of the function f at the point (x_0, y_0) are defined as

$$f_x(x_0, y_0) = \lim_{\Delta \to 0} \frac{f(x_0 + \Delta, y_0) - f(x_0, y_0)}{\Delta},$$

$$f_y(x_0, y_0) = \lim_{\Delta \to 0} \frac{f(x_0, y_0 + \Delta) - f(x_0, y_0)}{\Delta}.$$

If these definitions look intimidating, take a closer look, say, at the first one. Since the y-value is fixed, the numerator on the right-hand side depends only on one variable (a situation familiar from univariate calculus). Similarly, in the second formula, the x-variable is fixed. This observation suggests the following rule for calculating partial derivatives of a function of two variables: *When you differentiate with respect to one of the two variables, treat the other as a constant.*[3]

Fact 2: Necessary condition for extremum of a function of two variables. If a function $f(x, y)$ attains a local extremum (minimum or maximum) at a point (x_0, y_0), then (x_0, y_0) is a critical point. **Critical points** of a function of two variables are defined as the points where both partial derivatives equal zero or at least one of them does not exist.

Our function of interest $g(a, b)$ has partial derivatives at any point on the plane, so we are looking for the points where both first partial derivatives of this function equal zero.

Example 10.15. Derive the system for finding the optimal parameters a^*, b^* that define the line of best fit for the data $[(x_k, y_k)]_{k=1}^n$.

Solution. First, we calculate the partial derivatives of g and equate them to zero.

$$g_a'(a, b) = 0 = 2 \sum_{k=1}^n x_k \cdot (a x_k + b - y_k),$$

$$g_b'(a, b) = 2 \sum_{k=1}^n (a x_k + b - y_k) = 0.$$

With notation

$$C_1 \equiv \sum_{k=1}^n x_k^2, \quad C_2 \equiv \sum_{k=1}^n x_k, \quad C_3 \equiv \sum_{k=1}^n x_k y_k, \quad C_4 \equiv \sum_{k=1}^n y_k,$$

these equations can be rewritten as a system of two linear equations in two unknowns:

$$C_1 a + C_2 b = C_3, \quad C_2 a + n b = C_4. \tag{10.8}$$

It can be shown that the determinant of the system (10.8) is nonzero. Therefore, the solution to this system is the unique critical point of the function g. Proving that the function attains the global minimum at this point involves more tools of multivariable calculus, and we will take this fact for granted.

Exercise 10.16.

---

[3]The rule actually applies to functions of any number of variables.

(a) Manually find the line of best fit for the data $[(-5, -8), (4, 2), (9, 7)]$.

(b) Make a CAS function **myLS(L)** that takes a list of data $L = [(x_k, y_k)]_{k=1}^n$, solves the linear system (10.8), and returns

  - the equation of the line of best fit
  - the plot of the data and the line of best fit in one figure.

Test your function on the data in part (a).

Derivation of the system (10.8) **in matrix form (optional).** We show this derivation using only two data points. The solution to this linear LS problem is known up front – it is the line passing through the two data points – but let us pretend that we do not know this.

Consider two data points $[(x_1, y_1), (x_2, y_2)]$. The error function (10.7) can be seen as the square of the magnitude of the vector

$$\mathbf{v} = \left[\begin{array}{c} a x_1 + b - y_1 \\ a x_2 + b - y_2 \end{array} \right].$$

Introduce the notation

$$X = \left[\begin{array}{cc} x_1 & 1 \\ x_2 & 1 \end{array} \right], \, \beta = \left[\begin{array}{c} a \\ b \end{array} \right], \, \mathbf{y} = \left[\begin{array}{c} y_1 \\ y_2 \end{array} \right].$$

☼ **Check Your Understanding.** Show that $\mathbf{v} = X\beta - \mathbf{y}$.

With this notation, the function (10.7) can be written as[4]

$$\mathbf{v}^T\mathbf{v} = \beta^T X^T X \beta - 2\beta^T X^T \mathbf{y} + \mathbf{y}^T \mathbf{y}.$$

It can be shown that the vector of partial derivatives of the function g equated to zero yields the so called **normal equation**:

$$X^T X \beta = X^T \mathbf{y}. \tag{10.9}$$

The solution to the normal equation is

$$\beta = (X^T X)^{-1} X^T \mathbf{y}. \tag{10.10}$$

Exercise 10.17.

(a) Show that for two data points the normal equation (10.9) is equivalent to the system (10.8).

(b) Show that for the data $[(1, 1), (3, 5)]$, equation (10.10) gives the coefficients of the equation of the line passing through the data points.

📖 It is known that the normal equation (10.10) for the parameters of the line of best fit works for any finite set of data points (with an appropriate modification of the definitions of X, β, and $\mathbf{v}$).

---

[4]The transposition property $(AB)^T = B^T A^T$ used here can be easily verified for 2×2 matrices.

10.5 Two real-life applications of the LS method

In some cases, when there is no linear association in the given data, certain knowledge about the relation between the predictor and response quantities can be used to reduce the modeling problem to a linear LS problem. For instance, there is ample evidence that some properties of many complex systems can be described by the power law $f(x) = c x^r$. It is desirable to estimate the parameters c, r based on available data. For positive values of the predictor and response variables that are presumably related by a power law, reduction to a linear LS problem is possible. In fact, then $\ln(f(x)) = r \ln(x) + \ln(c)$, and the linear LS applied to the data $L = [(\ln(x_k), \ln(f_k))]_{k=1}^n$ can be used to approximate $\ln(c)$ and r. Once $\ln(c)$ is found, calculating the parameter c is elementary. This version of the LS method is called a **logarithmic fit**. A log-log plot is a two-dimensional graph of numerical data that uses a logarithmic scale on both the horizontal and vertical axes. The data L is called **log-log data**. Such a modification of the LS method will be used to approximate relationships between the total body mass and skeleton mass of mammals in the next lab.

10.5.1 Lab 13: Bone weight vs weight for mammals.

Problem formulation. Zoologists have measured the skeletal bone weight and body weight (in kilograms) for various mammals. This data is summarized in the following table:[5]

animal	shrew	mouse	cat	rabbit	beaver	human	elephant
weight (kg)	0.0063	0.0295	0.845	2.0	22.7	67.3	6600
bone weight (kg)	0.0003	0.0013	0.0436	0.181	1.15	12.2	1782

It is commonly accepted that these two quantities are related by a power law. Using the data provided in the table above and the LS method, approximate the power law $BoneWeight = c(Weight)^r$ for mammals through the following steps.

Part 1. Define the lists w and bw for the weight values and bone weight values, respectively. Plot the points (w_j, bw_j), $j = 0, \ldots, 6$. Do you think that there is a linear trend in the scatterplot? Explain.

Part 2. Make a list of log-log points and plot them. Comment on what you see.

Part 3. Use an appropriate CAS fitting command to find the line of best fit $\ln(bw) = a \ln(w) + b$ for the log-log data. Superimpose the line on the log-log plot of the data. Your figure will be similar to Fig. 10.7.

Part 4. Write a power law relating weight and bone weight of mammals. Test your function to estimate the bone weight of a person who weighs 86 kg. (Answer: about 10.8 kg.)

Part 5. A relation between the total body mass and skeleton mass can be used to estimate the total mass of extinct mammals. Suppose that a skeleton of an extinct mammal weighs 15 kg. Use the found power law to estimate the total body weight of the mammal.

In the last lab of this chapter several approximations between two physical quantities, the restoring force and displacement of a rubber band, are constructed and compared.

---

[5]The data given in the table is from a problem in [**8**].

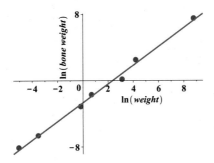

Figure 10.7. The log-log data and the line of best fit for the problem in the Lab 13 "Bone weight vs weight for mammals".

10.5.2 Lab 14: Modeling relationship between the restoring forces and displacements.
In this lab the relation between the displacement of an elastic rubber band and restoring forces is modeled using both linear and nonlinear LS methods. The nonlinear version of the method uses a pool $\mathcal{F}$ of functions $f(x, \alpha)$ that are nonlinear in the variable x and **depend linearly on the parameters** $\alpha = [\alpha_1, \alpha_2, \ldots, \alpha_n]$. For example, if we choose a family of cubic polynomials $f(x, \alpha) = x^3\alpha_3 + x^2\alpha_2 + x\alpha_1 + \alpha_0$ as models for fitting some data $[(x_k, y_k), k = 1 \ldots n]$, then we have to minimize the function (10.6) depending on four parameters α_j, $j = 0, 1, 2, 3$.

We know from everyday practice that there is a relation between the force needed to stretch or compress, say, an elastic band by some distance. In physics, the relation between the stretching/compressing force F and the displacement Δ (change in position) is described in the linear case by Hooke's law, $F = \pm k \cdot \Delta$, where the coefficient of proportionality k characterizes the stiffness of the band. For displacements larger than a certain value specific to each elastic material, Hooke's law is no longer valid and the relation becomes more complicated. This relation is extremely important in applied sciences, in particular, in engineering.

Problem formulation. Consider an elastic rubber band stretched with different forces. Corresponding displacements Δ and the stretching forces F are shown in the table below.

Δ (m)	.01	.02	.03	.05	.06	.08	.10	.13	.16	.18	.21	.25	.28
F (N)	0.21	0.42	0.63	0.83	1.0	1.3	1.5	1.7	1.9	2.1	2.3	2.5	2.7

Do the following.

Part 1. Make a CAS function **compareLS(L, p)** that takes a list L of data points and parameter p and returns

- the linear LS fit if $p = 1$
- the quadratic LS fit if $p = 2$
- the logarithmic fit $y = a + b\ln(x)$ if $p = 3$

Use an appropriate curve-fitting CAS command for this task. Your function should also return the sum of squares of corresponding residuals.[6] Notice that the LS method for a quadratic fit involves solving a system of three linear equations.

Part 2. Use the function made in Part 1 to find all three LS fits for the given data. In each case superimpose the approximating curve on the data plot.

Part 3. In each case calculate the average of the predicted force values and compare it with the average of the measured forces. Comment on the results of your computations.

10.6 Glossary

- n^{th} **Taylor polynomial** of a function f:

$$P_n(x) = f(c) + f'(c)(x - c) + \frac{f''(c)}{2!}(x - c) \cdots + \frac{f^{(n)}}{n!}(c)(x - c)^n.$$

The accuracy of the approximation of the function f by its Taylor polynomial can be estimated by the absolute value of the **remainder term**

$$R_n(x) = \frac{f^{(n+1)}(\xi)}{(n + 1)!}(x - c)^{n+1},$$

where ξ is some number between x and c.

- **Lagrange polynomial interpolant** of degree n – For a function f with **nodal values** f_k at the **nodes** x_k, $k = 0, \ldots, n$, the Lagrange interpolant is

$$L_n(x) = \sum_{k=0}^{n} f_k l_k(x),$$

where $\{l_k(x)\}_{k=0}^{n}$ is the **Lagrange basis** for the given set of nodes.

- **Spline interpolant** – A special type of continuous **piecewise polynomial**. The highest degree of a spline interpolant is called **the degree of the spline**. **Cubic splines** are twice differentiable functions.

- **Regression line (line of best fit)** for a data set $\{(x_k, y_k)\}_{k=1}^{n}$ – The line $y = a^*x + b^*$, such that the parameters a^*, b^* minimize the sum of the squares of the **residuals**

$$g(a, b) = \sum_{k=1}^{n}(a x_k + b - y_k)^2.$$

- **Least squares method** – A data fitting method used to find a function in a set of specified functions $f(c, x)$ depending on k parameters $c = (c_1, c_2, \ldots, c_k)$ that minimises the sum of the squares of the residuals $f(c, x_k) - y_k$ for a set of data $\{(x_k, y_k)\}_{k=1}^{n}$.

---

[6]See Definition 10.13.

11

Trigonometric Approximation

We saw that a Taylor polynomial approximation of a function is a good model only in some neighborhood of its center. Another restriction is that the function of interest must be differentiable some number of times. Lagrange polynomial interpolation requires the function of interest to be continuous, which could be too restrictive in some applications. Over the years, some mathematical tools for representing functions that may have discontinuities have been developed. A Fourier series is a prominent example of such a tool.

The Fourier series of a periodic function is an infinite sum of sines and cosines with coefficients defined by the function. For a 2π-periodic function the series is of the form

$$S(t) = \frac{a_0}{2} + \sum_{k=1}^{\infty} a_k \cos(k\,t) + b_k \sin(k\,t).$$

Any partial sum of a Fourier series is an example of a so called **trigonometric polynomial**. The partial sums can be used for approximating periodic functions satisfying fairly general assumptions. For example, consider the red curve shown in Fig. 11.1. This discontinuous periodic signal called a **square wave** is a common function that appears in applications. The blue curve in this figure is an approximation of the square wave by a trigonometric polynomial with just a few terms. Partial sums of the Fourier series of a periodic function have certain advantages over other trigonometric polynomials in approximating the function. In this chapter we will use the term "trigonometric polynomials" only for the partial sums of Fourier series.

After a short review of trigonometric functions, we will illustrate properties of trigonometric polynomial approximations and their applications in examples. The formulas for Fourier coefficients will be derived, and partial sums of Fourier series will be used in examples for approximating **periodic piecewise differentiable functions**. The chapter includes two mini projects, and a lab. The goal of the lab is to compare the accuracy of trigonometric polynomial approximations of two periodic functions (one continuous and the other piecewise continuous). The first mini project is on evaluating the famous Riemann function at a specific input. The second mini project is a

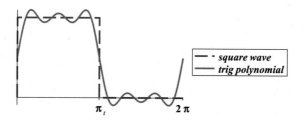

Figure 11.1. A square wave (red graph) approximated by trigonometric polynomial S_3 (blue curve) with only three terms.

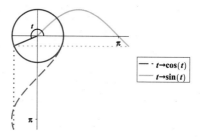

Figure 11.2. A snapshot of cosine and sine graphs generated as coordinates of the point moving counterclockwise along the unit circle.

classic solution of the heat flow problem in a thin circular ring. Fourier analysis of this problem was initiated by Jean-Baptiste Joseph Fourier, the father of Fourier series.

For simplicity, only 2π-periodic functions are considered. With some simple algebraic adjustments, Fourier analysis can be extended to periodic functions with any period, which we touch upon later in Chapter 12.

11.1 Short review of trigonometric functions

Trigonometric functions are introduced as functions of an angular argument. We will assume that all angles are measured in radians. The length of an arc of the unit circle is numerically equal to the measurement in radians of the angle that it subtends. Since the circumference of the unit circle is 2π, the radian measure of one complete revolution is 2π, and one radian is just about $360/(2\pi) \approx 57.3°$.

Exercise 11.1. Make a function ***convert_angle*(t, p)** that takes an angle t and a parameter p. The value $p = 0$ indicates that the angle t is in degrees, and the function should return the value of the angle in radians. The value $p = 1$ signals that t is in radians, and the function should return the measure of this angle in degrees. Test your function on an example of your choice.

Recall that for angles $t \in (0, \pi/2)$, the functions $\sin(t)$ and $\cos(t)$ can be defined as ratios of sides of a right triangle with an acute angle t. The definition of these functions is extended to an arbitrary angle using the unit circle centered at the origin. Given any angle t, there is a unique point on the unit circle that corresponds to it, and $\cos(t)$, $\sin(t)$ are defined as the x- and y-coordinates, respectively, of this point. Fig. 11.2 shows a snapshot of creating the graphs of sine and cosine using the radius-vector of a point running along the unit circle in the counterclockwise direction. It follows from this definition that the sine function is odd and the cosine function is even. The function $\tan(t)$ is defined as the ratio of sine and cosine. It is a π-periodic function, and its domain does not include the points $\pi/2 + k\pi$ with $k \in \mathbb{Z}$ (where $\cos(t) = 0$). Fig. 11.3(a) shows the graph of the function $\tan(t)$ and the vertical asymptotes $x = -\pi/2$ and $x = \pi/2$. One can interpret the values of the tangent function using the unit circle as follows. Draw the tangent line to the circle passing through the point $(1, 0)$ and extend out the radius that defines the angle of interest t until it hits the tangent. The y-value of the intersection point is $\tan(t)$. See Fig. 11.3(b).

💡 **Check Your Understanding.** Explain why the y-value of the intersection point is $\tan(t)$.

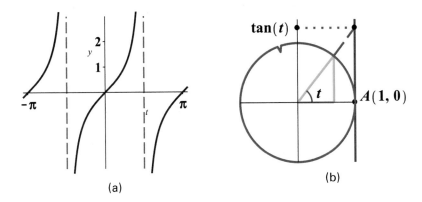

Figure 11.3. (a) The graph of $\tan(t)$. (b) Geometric interpretation of the tangent function values.

Exercise 11.2. Using Euler's formula, we can express the sine and cosine functions in terms of the complex exponential function. Use these representations and appropriate algebraic manipulations to derive the trigonometric identities below.

- $2 \sin(\alpha) \sin(\beta) = \cos(\alpha - \beta) - \cos(\alpha + \beta)$,

- $2 \cos(\alpha) \cos(\beta) = \cos(\alpha + \beta) + \cos(\alpha - \beta)$,

- $2 \sin(\alpha) \cos(\beta) = \sin(\alpha + \beta) + \sin(\alpha - \beta)$.

A function f is **periodic** if there exists a nonzero constant $P > 0$ such that $f(t + P) = f(x)$ for all x in the function domain. The period of f is the smallest of such P-values. The function $\sin(t)$ has the period 2π. It follows that the period of the function $\sin(\omega t)$ is $2\pi/\omega$ (**think why**). The **frequency** of a periodic function is the number of oscillations (**cycles**) per unit time. The SI unit for frequency is called the **hertz** (Hz). Mathematically, frequency is the reciprocal of the period. For example, the frequency of $\sin(t)$ is $1/(2\pi) \approx 0.159$ Hz.

The **amplitude** of a periodic function is the maximum displacement from its mean value. The amplitude of the sine and cosine functions is one. A periodic function $S(t) = a\cos(kt) + b\sin(kt)$, $a, b \in \mathbb{R}$, can be written as $S(t) = A((a/A)\sin(kt) + (b/A)\cos(kt))$ with $A = \sqrt{a^2 + b^2}$. Let $\varphi \in (-\pi, \pi]$ be the unique angle with $\sin(\varphi) = a/A$ and $\cos(\varphi) = b/A$. Then

$$S(t) = a\cos(kt) + b\sin(kt) = A\cos(kt - \varphi).$$

Here, we used the trig identity $\cos(\alpha)\cos(\beta) + \sin(\alpha)\sin(\beta) = \cos(\alpha - \beta)$. Thus, the amplitude of S is A. The angle φ is called the **phase** of the function S.

📖 In mechanics and physics, the linear combination $S(t) = a\sin(kt) + b\cos(kt)$ describes what is called **simple harmonic motion.**[1] In acoustics, these functions are called harmonics, or harmonic waves. The harmonic functions in higher dimensions are defined as solutions to a certain partial differential equation (PDE). They constitute the main object of study in the branch of mathematics called **harmonic analysis**.

Exercise 11.3. Make a CAS function *convert_harmonic*(**k, a, b**) that takes the parameter k and coefficients a, b of a harmonic $S(t) = a\sin(kt) + b\cos(kt)$ and returns its amplitude and phase. Test your function on $S(t) = 3\sin(t) + 4\cos(t)$. **Answer:** $A = 5$, $\varphi = \arccos(0.8) \approx 0.64$, or $\approx 36.9°$.

Some real-life examples of periodic functions. Various important real-life periodic processes can be modeled by harmonics. Three examples are given below.

Sound waves. Humans perceive periodic waves through a fluid medium, such as air or water, as sounds. The human ear can detect such waves with frequencies ranging between roughly 20 Hz to 20 kHz. The intensity of a sound, as perceived by human ears, is determined by the energy of the sound wave motion in the medium. At any given point in the medium the **energy content** of the wave disturbance varies as the **square of the amplitude** of the wave motion. That is, if the amplitude of the oscillation is doubled, the energy of the wave motion is quadrupled.

🔦 *Check Your Understanding.* Calculate the frequency of the harmonics $S_1(t) = \sin(2\pi t)$ and $S_2(t) = \sin(0.2\pi t)$. Plot the function graphs in one figure over the longest period of the two functions. Can a human hear the sound wave represented by the function $S(t) = \sin(2\pi t)$?

Exercise 11.4. A common concern in audio engineering is the so called **phase cancellation** phenomenon. If two identical signals are out of phase then they can potentially cancel each other out. For example, placing two microphones incorrectly can cause this issue.

(a) Consider two signals represented by the functions $S_1(t) = 10\sin(t - \frac{\pi}{2})$ and $S_2(t) = 10\sin(t - \frac{3\pi}{2})$. Use trigonometric identities to show that the combined signal $S_1(t) + S_2(t)$ has amplitude zero (i.e., S_1 and S_2 cancel one another).

---

[1]This two-parameter family of functions is the general solution to the ordinary differential equation $x'' + k^2 x = 0$.

Figure 11.4. Beats phenomenon.

(b) Consider the signals $S_1(t, \beta_1) = A \sin(t - \beta_1)$ and $S_2(t, \beta_2) = A \sin(t - \beta_2)$. Determine the relation between the phases β_1 and β_2 such that the signals cancel each other out. That is, $S_1(t, \beta_1) + S_2(t, \beta_2) = 0$.

Beats. A beat is the superposition (sum, interference) of two harmonics with slightly different frequencies. It is perceived by human ears as periodic variations in sound intensity. The rate of the intensity variations is defined by the difference between the two frequencies. We experience this effect when tuning a musical instrument to match the pitch of a tuning fork. We hear a characteristic "wa-wa" sound as the two frequencies approach each other.

Example 11.5. Fig. 11.4 shows the graph of the sum of two sine waves with close frequencies, $\sin(2\pi t)$ and $\sin(1.6\pi t)$. Using the appropriate formula from Exercise 11.2, we obtain $\sin(2\pi t) + \sin(1.6\pi t) = 2 \sin(0.2\pi t) \cos(1.8\pi t)$. Two green plots are added to Fig. 11.4 to show the periodic change of the amplitude $2 \sin(0.2\pi t)$ of the sum.

Exercise 11.6. Convert the sum $\cos(3\pi t) + \cos(2.7\pi t)$ into a product of trigonometric functions and make a figure similar to Fig. 11.4.

Resonance. For any system there exists a specific frequency called its **natural frequency**. This is the rate at which a system oscillates in the absence of any driving or damping force. When an external periodic force whose frequency matches the natural frequency is applied, the amplitude of the system's vibrations grows. This phenomenon is called **resonance**.

A familiar example of resonance is a playground swing. One system is a swing and person, say a child sitting in a hanging seat, and the other is a periodic force exerted by some external agent (a person or a device) on the swing. Resonance can also be observed in the motion of a forced mass-spring oscillator when the frequency of the external force is close to the natural frequency of the oscillator. This is a typical topic in differential equations since the motion of a forced mass-spring oscillator can be modeled by a certain ordinary differential equation.

Resonance can be desirable if you want to amplify oscillations, such as optimizing acoustics in a concert hall. However, large oscillations can be dangerous and can even lead to the destruction of buildings and bridges. (For a real example of such an unfortunate event google "The Millennium Bridge collapse".)

📖 Trigonometric functions are defined using the unit circle, which is why they are also called "circular functions". There are functions called "**elliptic functions**" that are related to ellipses through so called **elliptic integrals**. (Originally, elliptic integrals occurred when calculating the arc length of an ellipse.) Elliptic functions appear as solutions to many important problems in classical mechanics. The theory of elliptic functions was developed in the 19th century by Gauss, Abel, Jacobi, and their contemporaries.

11.2 Fourier series

A Fourier series gives a representation of a 2π-periodic function f as a sum of functions from the **standard trigonometric set of basis functions**

$$\{1, \sin(k\,t), \cos(k\,t), \ k \geq 1.\}$$

It has the form

$$S(t) = \frac{a_0}{2} + \sum_{k=1}^{\infty} a_k \cos(k\,t) + b_k \sin(k\,t) \tag{11.1}$$

with coefficients defined by the formulas

$$a_k = \frac{1}{\pi} \int_{-\pi}^{\pi} f(t) \cos(k\,t) d\,t, \ k \geq 0, \tag{11.2a}$$

$$b_k = \frac{1}{\pi} \int_{-\pi}^{\pi} f(t) \sin(k\,t) d\,t, \ k \geq 1. \tag{11.2b}$$

In the language used in applications, a Fourier series is a decomposition of a periodic signal into the sum of sine and cosine waves. This decomposition reveals the frequency content of the signal.

11.2.1 Deriving formulas for Fourier coefficients. Consider a piecewise differentiable 2π-periodic function f. It can be shown that

- If f is continuous at a point t, the sum of the series (11.1) with coefficients (11.2) equals $f(t)$

$$f(t) = \frac{a_0}{2} + \sum_{k=1}^{\infty} a_k \cos(k\,t) + b_k \sin(k\,t). \tag{11.3}$$

- If t is a jump discontinuity of the function, the sum of the series (11.1) is $\frac{1}{2}[f(t^-) + f(t^+)]$, where $f(t^{\pm})$ are one-sided limits of f at the point t.

- The sum of the series obtained by term-by-term integration of the right-hand side of equation (11.3) over any interval equals the integral of f over the same interval.

To begin deriving the formulas in (11.2), let us start with computing the following integrals:

$$J_1 = \int_{-\pi}^{\pi} \sin(k\,t) \sin(m\,t) d\,t,$$

$$J_2 = \int_{-\pi}^{\pi} \sin(k\,t) \cos(m\,t) d\,t,$$

$$J_3 = \int_{-\pi}^{\pi} \cos(k\,t) \cos(m\,t) d\,t.$$

Exercise 11.7.

(a) Show that $J_1 = J_3 = 0$ for $k \neq m$ and $J_2 = 0$ for any k and m.

   Hint: Use the formulas derived in Exercise 11.2.

(b) Show that $J_1 = J_3 = \pi$ for $k = m$.

Next, multiply equation (11.3) by $\cos(mt)$ and integrate both sides[2] over the interval $[-\pi, \pi]$. For $m \neq 0$, use the formulas for the integrals J_i, $i = 2, 3$, to obtain

$$\int_{-\pi}^{\pi} f(t)\cos(mt)dt = \frac{a_0}{2}\int_{-\pi}^{\pi} \cos(mt)dt + \pi a_m = 0 + \pi a_m,$$

and formula (11.2a) follows for $k > 0$. The formula also applies for calculating a_0. This explains why the form $(1/2)a_0$ is chosen for the constant term of the Fourier series.

☀ **Check Your Understanding.**

(a) Show that the formula (11.2a) also applies for calculating a_0.

(b) In a similar fashion derive formula (11.2b).

Remark 11.8. It can be shown that choosing any interval of integration of length 2π for the formulas in (11.2) will produce the same Fourier coefficients for a piecewise differentiable 2π-periodic function f.

☀ **Check Your Understanding.** Justify the following statements.

(a) If f is an odd function, then $a_k = 0$ for all $k \geq 0$, and $b_k = \frac{2}{\pi}\int_0^{\pi} f(t)\sin(kt)dt$ with $k \geq 1$.

(b) If f is even, then $b_k = 0$ for all $k \geq 1$, and $a_k = \frac{2}{\pi}\int_0^{\pi} f(t)\cos(kt)dt$, $k \geq 0$.

Example 11.9.

(a) Find ten Fourier coefficients of the 2π-periodic function defined on the interval $(-\pi, \pi]$ as $f(t) = t$.

(b) Identify the pattern in the Fourier coefficients found in part (a) and write the Fourier series $S(t)$ of the function f.

(c) Make two figures, one with the plot of f and the partial sum S_3 of S, and the other with the graphs of f and the partial sum S_5.

Solution.

(a) The given function is odd. Therefore we need to find only the coefficients b_k. With CAS assistance, we obtain the following sequence of ten b-coefficients:

$$(2, -1, 2/3, -1/2, 2/5, -1/3, 2/7, -1/4, 2/9, -1/5).$$

---

[2]The right-hand side is integrated term-by-term.

(b) After some visual inspection and contemplation, we conclude that

$$b_k = \frac{(-1)^{k-1} \cdot 2}{k} \Rightarrow f(t) = 2 \sum_{k=1}^{\infty} \frac{(-1)^{k-1}}{k} \sin(k\,t).$$

(c) $S_3(t) = 2\sin(t) - \sin(2\,t) + \frac{2}{3}\sin(3\,t);\ S_5(t) = S_3(t) - \frac{1}{2}\sin(4\,t) + \frac{2}{5}\sin(5\,t).$

Fig. 11.5 shows the graphs of the given function and the two trigonometric approximations we constructed.

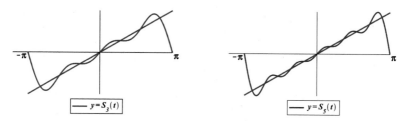

Figure 11.5. Two trigonometric approximations of the function in Example 11.9.

From Fig. 11.5 one can notice that the partial Fourier sum approximations overshoot near the points of discontinuity of the function. This is a general effect known as **Gibbs phenomenon**. It is undesirable in signal processing and certain mathematical techniques have been developed to reduce the effect of Gibbs phenomenon.

The goal of the next exercise is to create two CAS help functions to assist in calculating Fourier coefficients.

Exercise 11.10.

(a) Make a CAS function *acoef*(**fcn**, **n**) that takes a piecewise differentiable 2π-periodic function and a positive integer n, and returns a list of $n + 1$ Fourier coefficients $a_k,\ k = 0, \ldots, n$, of the function.

(b) Make a CAS function *bcoef*(**fcn**, **n**) that takes a piecewise differentiable 2π-periodic function and a positive integer n, and returns a list of n Fourier coefficients $b_k,\ k = 1, \ldots, n$, of the function.

In the next exercise, use an appropriate help function to find a trigonometric approximation of the given function.

Exercise 11.11. Consider the 2π-periodic function $f(x) = |x|, x \in (-\pi.\pi]$, called the **triangular wave**.

(a) Find three partial sums S_n, $n = 3, 4, 6$, of the Fourier series of this function.

(b) Make three figures, each with a graph of f and a graph of one of the trigonometric polynomial approximations S_n, $n = 3, 4, 6$, over three periods of the given function.

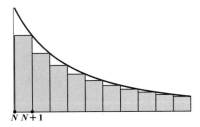

Figure 11.6. Visualization of the remainder R_N of the p-series with $p = 2$. The curve is the graph of $1/x^2$.

11.3 About the accuracy of trigonometric approximations

Infinite functional series provide mathematical tools to exactly represent functions that satisfy certain requirements. However, in practice, we can deal only with partial sums. It is important to have a means for estimating an approximation error, the remainder R_N, when you keep only N terms of a series representation. Different convergence types of Fourier series are studied using advanced methods of analysis. We need only the following simple estimate of the remainder for a piecewise differentiable 2π-periodic function:

$$|R_N| = |f(t) - S_N(t)| \leq \sum_{N+1}^{\infty} (|a_k| + |b_k|),$$

where $S_N = a_0/2 + \sum_{k=1}^{N} a_k \cos(k t) + b_k \sin(k t)$. We will also use the following estimate of the remainder for one of the p-series. (Recall that a p-series is defined as $\sum_{n=1}^{\infty} 1/n^p$.)

Claim. The remainder R_N of the 2-series $\sum_{n=1}^{\infty} 1/n^2$ is less than $1/N$.

Proof. In Fig. 11.6 the remainder $R_N = \sum_{N+1}^{\infty} 1/n^2$ is depicted as the sum of the areas of the green rectangles with base one. Inspecting the figure, we obtain the estimate

$$R_N < \frac{1}{(N+1)^2} + \int_{N+1}^{\infty} 1/x^2 \, dx = \frac{1}{(N+1)^2} + \frac{1}{N+1}.$$

$\square$

💡 **Check Your Understanding.** Calculate the difference between $1/N$ and the right-hand side of this equation to complete the proof.

In the next example we will put this claim to work.

Example 11.12. Consider the 2π-periodic function f given on the interval $(-\pi, \pi]$ by the equation $f(t) = t^2$. Find an estimate for $N \in \mathbb{N}$ such that $|f(t) - S_n(t)| \leq 0.1$ for all $t \in \mathbb{R}$ and $n \geq N$.

Solution. Using CAS assistance, we find that the Fourier series for the function f has the form

$$\frac{\pi^2}{3} - 4\left(\cos(t) + \frac{\cos(2 t)}{2} + \cdots + \frac{\cos(k t)}{k^2} + \cdots \right).$$

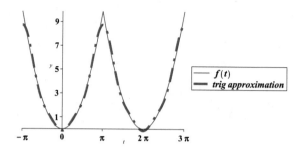

Figure 11.7. The remainder plots for Example 11.12.

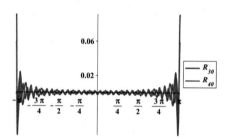

Figure 11.8. The remainder plots for Example 11.12.

The required accuracy will be achieved for the Nth partial sum of the Fourier series with N satisfying the inequality

$$4 \left| \sum_{N+1}^{\infty} \frac{\cos(k\,t)}{k^2} \right| < 0.1.$$

Claim 1 implies that the sum is bounded above by $1/N$. The inequality $4/N \leq 0.1$ yields $N \geq 40$.

It may seem unexpected that so many terms are needed to meet a rather modest accuracy requirement. Fig. 11.7 shows a sizable deviation of the trig approximation from the given function near the points $\pm(2m-1)\pi$, $m \in \mathbb{N}$, where the function is not differentiable. Fig. 11.8 suggests a somewhat quantitative demonstration: at these points we see significantly larger values of the remainders. Therefore, more terms are needed to meet the accuracy requirement for **all values** of t.

The sum of the p-series with $p = 2$ is the value of the Riemann zeta function $\zeta(s) = \sum_{n=1}^{\infty} 1/n^s$ at $s = 2$. In the next mini project you will find this value **exactly** using a certain Fourier series.

11.3.1 Mini project: Using Fourier series for finding a value of the Riemann zeta function. Consider the 2π-periodic function f defined for $t \in [0, 2\pi)$ as $f(t) = t^2/4 - \pi t/2$. Find the Fourier series of this function and use it to find $\zeta(2)$.

Suggested directions.

Step 1. Prove that f is an even function. Is f continuous? Explain.

Step 2. Find ten Fourier coefficients for the function f.

Step 3. Identify the pattern in the Fourier coefficients and write the Fourier series for f.

Step 4. Use the Fourier series found in Step 3 to find $\zeta(2) = \sum_{n=1}^{\infty} 1/n^2$. **Hint**: Recall the relation between the values of a continuous periodic function and the sum of its Fourier series.

11.3.2 Root mean square. In the lab below you will use the so called **root mean square (rms)** of the remainders as a measure of the trigonometric polynomial approximation error. Here is the definition of this concept.

Definition 11.13. The root mean square of a $2T$-periodic function g is

$$\mathbf{rms}(g) = \sqrt{\frac{1}{2T}\int_{-T}^{T} |g(t)|^2 dt}.$$

The radicand is called the **mean square**.

Check Your Understanding. Consider the function $f(t) = A\sin(kt)$. Show that $\mathbf{rms}(f) = |A|/\sqrt{2}$.

In this chapter we are dealing with the set of standard trigonometric basis functions. The **rms** for any of these functions (except the constant 1) is $1/\sqrt{2}$. The mean square of a sum of functions from this set equals the sum of mean squares of these functions. The **rms** is the square root of this sum.

Example 11.14. Estimate the **rms** of the remainder R_3 for the Fourier series of the function defined in Example 11.9.
Solution. The Fourier series found in Example 11.9 is $2\sum_{k=1}^{\infty}(-1)^{k-1}\sin(kt)/k$. Since the **rms** of each term in the sum is $1/(k\sqrt{2})$, we have

$$\mathbf{rms}(R_3) = \sqrt{2\sum_{k=4}^{\infty} 1/k^2}.$$

This evaluates to[3] $\sqrt{2(\pi^2/6 - (1 + 1/4 + 1/9))} \approx 0.75$.

11.3.3 Lab 15: Comparing the trigonometric approximation accuracy of two periodic waves.

Problem formulation. Consider the 2π-periodic function

$$g(t) = \begin{cases} \sin(t) & 0 \le t \le \pi, \\ 0 & \pi < t < 2\pi, \end{cases}$$

called in signal processing a "half-wave rectified sinusoid".

Part 1. Construct the Fourier series of the function g and define its partial sum S_3.

[3]We use the value $\zeta(2) = \pi^2/6$ here.

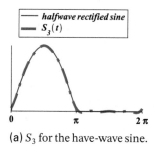

(a) S_3 for the have-wave sine.

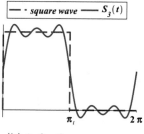

(b) S_3 for the square wave.

Figure 11.9. Trigonometric approximations of a continuous and a piecewise-continuous functions.

Part 2. Make a figure with plots of g and S_3. Use CAS assistance to estimate the **rms** of the remainder R_3.

Part 3. Repeat Parts 1 and 2 for the square wave

$$h(t) = \begin{cases} 1 & 0 \leq t \leq \pi, \\ 0 & \pi < t < 2.\pi \end{cases}$$

Part 4. Compare the **rms**(R_3) values for the remainders found in Parts 2 and Part 3 and comment on the results of this comparison. Note that the function g is continuous and the square wave is not.

11.4 Celebrated classical application of Fourier series

The motion of a free, undamped spring-mass system is modeled by certain ordinary differential equations. To model the oscillation of an extended object, such as a string or a thin rod, you need to find a function of two variables (time and position) that satisfies a certain **partial differential equation (PDE)**. Periodic variations of other physical quantities, such as the temperature in a rod or a two-dimensional region, are also modeled by PDEs.

Historically, the first application of Fourier series to partial differential equations was done by Fourier himself. He applied his invention, the Fourier series, to solve a PDE modeling heat flow in a thin circular ring. Before Fourier's work, no general solution to the heat equation was known.

In this section we state this problem in mathematical terms and derive its solution using Fourier series. To keep the presentation simple, we will not discuss any formal mathematical justification of the solution method for this problem. Fourier himself was not able to do this but for a different reason – such a justification was not known at his time. We also will not discuss here the physical aspects of the problem, such as the derivation of the governing PDE.

11.4.1 Problem formulation and general solution. Consider a unit wire circle as a wire rod of length 2π. Suppose that the initial temperature distribution of the rod is given by a continuous, piecewise differentiable, 2π-periodic function f. Find the temperature distribution $u(x,t)$ in the rod at time $t > 0$.

It is known from physics that one-dimensional heat transfer is governed by the **heat equation** $u_t = u_{xx}$. Thus, we can state the problem as

Find the solution $u(x, t)$ to the heat equation $u_t = u_{xx}$ satisfying the initial condition $u(x, 0) = f(x)$, $x \in [0, 2\pi]$, $t \geq 0$.

This is an initial value problem (IVP) for the heat equation.

Remark 11.15. The heat equation derived on physical grounds includes a coefficient on the right hand side of the equation which incorporates certain physical properties of the wire. To simplify mathematical manipulations, the coefficient can be eliminated without loss of generality by a certain scaling procedure.

Using the complex form of a Fourier series for solving this problem would be more elegant. However, we will stay within our comfort zone and work out the problem using the real form introduced in this chapter.

Solution sketch. We will seek the solution as the sum of Fourier series with respect to the space variable x, but with **coefficients depending on time**:

$$u(x, t) = \frac{a_0}{2} + \sum_{k=1}^{\infty} a_k(t) \cos(k x) + b_k(t) \sin(k x). \tag{11.4}$$

Our goal is to find the functions $a_k(t)$, $b_k(t)$. If we freeze the variable t for a moment, formula (11.4) becomes just a standard Fourier series representation of a univariate function of x. The main difference is that the function is not known, and we want to use Fourier series as a device for finding it. Using the formula for the Fourier coefficients a_k we obtain

$$a_k(t) = \frac{1}{\pi} \int_0^{2\pi} u(x, t) \cos(k x) d x.$$

Differentiating both sides with respect to t yields $a'_k(t) = \frac{1}{\pi} \int_0^{2\pi} u_t(x, t) \cos(k x) d x$. Using the heat equation to replace the time derivative u_t in the integrand with u_{xx}, we get

$$a'_k(t) = \frac{1}{\pi} \int_0^{2\pi} u_{xx}(x, t) \cos(k x) d x.$$

Now we integrate by parts twice to move the derivatives off of u. The result is a simple separable ODE for the coefficients $a_k(t)$:

$$a'_k(t) = -\frac{k^2}{\pi} \int_0^{2\pi} u(x, t) \cos(k x) d x = -k^2 a_k(t). \tag{11.5}$$

Solving this ODE, we find that

$$a_k(t) = a_k(0) \exp(-k^2 t).$$

🔆 *Check Your Understanding.* Derive the ODE (11.5). Carefully explain why the boundary terms in the integration by parts formula vanish.

A similar sequence of mathematical manipulations leads to the solution for the coefficients b_k:

$$b_k(t) = b_k(0) \exp(-k^2 t).$$

Now the assumed solution ansatz (11.4) takes the form

$$u(x,t) = \frac{a_0(0)}{2} + \sum_{k=1}^{\infty} \left(a_k(0)\cos(k\,x) + b_k(0)\sin(k\,x) \right) \exp(-k^2 t). \qquad (11.6)$$

If the Fourier series of the initial temperature distribution is

$$f(x) = \frac{A_0}{2} + \sum_{k=1}^{\infty} A_k \cos(k\,x) + B_k \sin(k\,x),$$

then, by the initial condition, $a_k(0) = A_k$ and $b_k(0) = B_k$. In summary, we obtained the solution to the IVP:

$$u(x,t) = \frac{A_0}{2} + \sum_{k=1}^{\infty} \left(A_k \cos(k\,x) + B_k \sin(k\,x) \right) \exp(-k^2 t). \qquad (11.7)$$

This remarkable formula reduces the task of solving the IVP for the heat equation in a thin circular ring to the textbook exercise of finding Fourier coefficients of the given function that defines the initial temperature distribution. **"There is nothing as practical as a good theory."**[4]

⬚̣ Check Your Understanding.

(a) Show that the coefficient $A_0/2$ is the average value of the initial temperature distribution.

(b) Find the limit of the solution (11.7) as $t \to \infty$.

11.4.2 Mini project: Solving an IVP for heat transfer in a thin circular wire.

Problem formulation.

Step 1. Solve the IVP for the PDE $u_t = u_{xx}$, $u(x,0) = \pi - |x|$, $x \in [-\pi.\pi]$. Plot the function $u(x,0)$ and two snapshots of your solution for nonzero time values in one figure. Your figure will be similar to Fig. 11.10.

Step 2. Make a CAS function ***heat_in_ring*(fcn)** that takes a continuous piecewise differentiable, 2π-periodic function fcn and returns the solution of the IVP for the heat equation with $u(x,0) = f(x)$. Test your function on the problem in Part 1.

Remark 11.16. It is known that solutions to the heat equation with a non-smooth initial function, as in Part 1 of this mini project, become smooth in practically no time. The reader can try using smaller and smaller values of t and plot snapshots of the solution found in Part 1 of the mini project to see that the discontinuity of f' at the origin is "smoothed away" by the solution $u(x,t)$ practically instantly.

---

[4]This maxim is attributed to Kurt Levin, American psychologist. See Hunt, D. E., *Beginning with ourselves: In practice, theory and human affairs*, Cambridge, MA: Brookline Books, 1987, p 4.

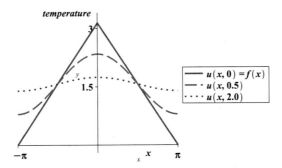

Figure 11.10. Snapshots of the temperature distribution in a thin circular ring with initial distribution defined by the function f.

11.5 Glossary

- **Periodic function** – A function f such that there exists a nonzero constant $P > 0$ such that $f(t + P) = f(x)$ for all x in the function domain. The period of f is the smallest of such P-values.

- **Frequency of a periodic function** – The number of oscillations (**cycles**) per unit time. The SI unit for frequency is called the **hertz** (Hz). Mathematically, frequency is the reciprocal of the period.

- **Fourier series** of a piecewise differentiable 2π-periodic function f:

$$S(t) = \frac{a_0}{2} + \sum_{k=1}^{\infty} a_k \cos(k t) + b_k \sin(k t),$$

with coefficients determined by f. The sum of the series equals the value of the function at all points where f is continuous, and to half of the sum of one-sided limits of f at the points of the jump discontinuities.

- **Harmonic wave (harmonic)** – The expression $a \cos(k t) + b \sin(k t)$. It can be written in the form $A \cos(k t - \varphi)$. The number A is the **amplitude** and φ the **phase** of the harmonic.

- **Mean square** of a $2T$-periodic function f:

$$\frac{1}{2T} \int_{-\pi}^{\pi} f^2(t)\,d t.$$

The root mean square, rms, of the function is the square root of its mean square.

12

Fourier Analysis in Music and Signal Processing

The goal of this chapter is to briefly explore how we can analyze and manipulate the frequency content of digital audio signals by transforming signals from the time domain to the so called "frequency domain". This is achieved through the use of a powerful mathematical tool called the **Fourier transform**. Using the concept of a Fourier series presented in the previous chapter, we will introduce and work through simple examples of the discrete and continuous Fourier transforms. This chapter's lab will then walk the reader through the steps of reading in an audio file and analyzing it using the Fourier transform. Finally, the last section of this chapter touches upon how Fourier series can be used to filter and compress a signal.

12.1 Introduction and background

As mentioned earlier, humans can distinguish pitches from about 20 Hz to 20 kHz. We also discussed that the function $S(t) = A \cos(k t - \phi)$ represents a sound wave with frequency $f = k/(2\pi)$ Hz and amplitude A, which is often measured using the decibel (dB) scale.[1] Such sound waves are called **pure tones**.

Example 12.1. Determine a function that represents the pure tone with frequency 1000 Hz.

Solution. We know that $f = k/2\pi = 1000$ Hz and so $k = 2\pi \cdot 1000 = 2000\pi$. Therefore, the function we are looking for is

$$S(t) = A \cos(2000\pi t).$$

So, if pure tones are represented mathematically by the sine and cosine functions, then how do we express other, more complex sounds? Remember that sounds are really

---

[1] The decibel is a commonly used unit in acoustics as a unit of sound pressure.

just longitudinal waves that move through a medium such as air or water. In physics, the **Principle of Superposition** states that when two or more waves overlap in space, the resultant wave is mathematically the algebraic sum of the individual waves. Therefore, this law implies that all sounds can be constructed from the building blocks of pure tones. That is, any sound is algebraically just a sum of pure tones with different amplitudes and frequencies.

Example 12.2. Two pure tones are being played at the same time. The first pure tone has an amplitude of 10 dB and frequency of 440 Hz. The second has an amplitude of 5 dB and frequency of 880 Hz. Write a function $S(t)$ that represents the combined sound wave of these two pure tones.

Solution. We know that the first pure tone can be represented by $S_1(t) = 10\cos(880\pi t)$ and the second by $S_2(t) = 5\cos(1760\pi t)$. By the superposition principle, the combined sound wave can be represented by the function

$$S(t) = S_1(t) + S_2(t) = 10\cos(880\pi t) + 5\cos(1760\pi t).$$

The individual pure tones and the combined signal are shown in Fig. 12.1.

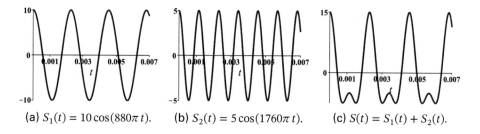

(a) $S_1(t) = 10\cos(880\pi t)$.   (b) $S_2(t) = 5\cos(1760\pi t)$.   (c) $S(t) = S_1(t) + S_2(t)$.

Figure 12.1. Plots of $S_1(t)$, $S_2(t)$, and $S(t) = S_1(t) + S_2(t)$.

The lowest frequency in a sum of pure tones is called the **fundamental frequency**. Humans perceive the fundamental frequency of a sound wave as the pitch, regardless of its amplitude. For example, the fundamental frequency in the previous example is 440 Hz.

Exercise 12.3. Consider the sound wave represented by the function

$$S(t) = 10\cos(200\pi t) + 20\cos(400\pi t) + 30\cos(800\pi t).$$

Identify the fundamental frequency of $S(t)$.

Frequencies greater than the fundamental frequency are called **harmonics** or **overtones**. For the signal $S(t)$ in the previous exercise, the overtones are 200 Hz and 400 Hz. All instruments and voices naturally produce specific overtones. The overtones present and their individual amplitudes are what makes every instrument sound unique, *even if they are playing the same pitch*. Fig. 12.2 shows fragments of waveforms produced by a guitar and piano playing the same note, middle A, which has a frequency of 440 Hz. The fundamental frequency of 440 Hz is the same in both waves, which is why we perceive the pitch as being the same for both instruments even though the instruments themselves sound very different.

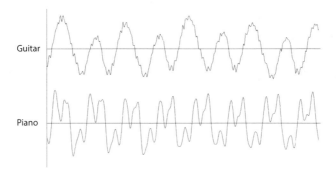

Figure 12.2. Waveforms produced by a guitar and piano playing the same note (A440).

The signals plotted in Fig. 12.1 and Fig. 12.2 are mathematical representations of continuous sounds in the **time domain**, meaning that the amplitude or **intensity** of the sound is expressed as a function of time. Representing a signal in the time domain is useful for various tasks:

- record and play back digital sound

- digitally increase or decrease the volume or intensity of the entire signal

- copy, paste, and delete portions of a signal

- signal compression

- digitally increase or decrease playback speed

On the other hand, many tasks require information about the specific frequencies that comprise a given signal. Some examples of such tasks are:

- pitch correction

- increasing/decreasing the intensity of specific frequencies (known as equalizing or EQ)

- analyzing frequency data

- removing certain frequencies (low/high pass filters)

The question now is how do we reveal the underlying frequency content of a signal from its time domain representation? We know by superposition that sounds are just sums of different pure tones. Therefore, if we can decompose a signal into its pure tones then we can easily identify which frequencies are present. This can be achieved through Fourier series.

12.2 Fourier series and periodic signals

Consider a differentiable, 2π-periodic signal $S(t)$. Recall from Section 11.2 that such a function has the following Fourier series representation:

$$S(t) = \frac{a_0}{2} + \sum_{k=1}^{\infty} a_k \cos(k\,t) + b_k \sin(k\,t). \tag{12.1}$$

We also mentioned that a harmonic wave $s(t) = a\cos(t) + b\sin(t)$ can be written as $s(t) = A\cos(t - \varphi)$ with amplitude $A = \sqrt{a^2 + b^2}$ and angle φ defined by conditions $\cos(\varphi) = a/A$, $\sin(\varphi) = b/A$. Applying this transformation to each harmonic wave in the series above allows us to rewrite the Fourier series as

$$S(t) = \frac{a_0}{2} + \sum_{k=1}^{\infty} A_k \cos(kt - \varphi_k).$$

In other words, the Fourier series representation of a 2π-periodic signal $S(t)$ effectively decomposes the signal into a sum of pure tones, revealing the frequency content of the signal. A visual representation of this idea is shown in Fig. 12.3. The red curve, which is the original signal, is effectively decomposed into a sum of pure tones at varying frequencies and amplitudes (blue curves).

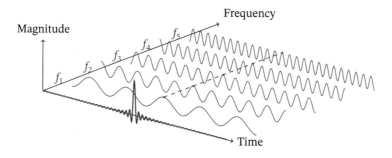

Figure 12.3. A visual representation of decomposing a signal (red curve) into a sum of pure tones at varying frequencies and amplitudes (blue curves).

The Fourier decomposition of a signal allows us to transform a signal from the time domain into the **frequency domain** or **spectrum** of a signal, where the intensity or amplitude of the signal is now given as a function of frequency, rather than time.

Example 12.4. In Example 12.2, we plotted the signal $S(t) = 10\cos(880\pi t) + 5\cos(1760\pi t)$ in the time domain (see Fig. 12.1). We will now plot the frequency content of this signal in the frequency domain. We know that the first pure tone in $S(t)$ has frequency 440 Hz and amplitude 10 dB. The second has frequency 880 Hz and amplitude 5. Using this information, we obtain the plot of $S(t)$ in the frequency domain shown in Figure 12.4.

Example 12.5. Let

$$S(t) = \begin{cases} 0 & -\pi \leq t \leq 0 \\ 1 & 0 < t \leq \pi \end{cases}.$$

As introduced in the previous chapter, this 2π-periodic function is known as a **square wave**. A plot of $S(t)$ in the time domain is shown below in Fig. 12.5a. Find the Fourier series decomposition of $S(t)$ and then use it to create a frequency domain plot.

Solution. As was done in Section 11.2, we will compute the Fourier series in the form of (12.1) by calculating the coefficients a_k and b_k using the formulas (11.2). First we

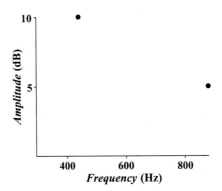

Figure 12.4. Plot of $S(t) = 10\cos(880\pi t) + 5\cos(1760\pi t)$ from Example 12.2 in the frequency domain.

calculate a_0:

$$a_0 = \frac{1}{\pi}\int_{-\pi}^{\pi} S(t)dt = \frac{1}{\pi}\int_0^{\pi} 1\,dt = \frac{1}{\pi}\cdot\pi = 1.$$

Now for $k \neq 0$,

$$a_k = \frac{1}{\pi}\int_{-\pi}^{\pi} S(t)\cos(kt)dt = 0 \tag{12.2}$$

and

$$b_k = \frac{1}{\pi}\int_{-\pi}^{\pi} S(t)\sin(kt)dt = \frac{1-\cos(\pi k)}{\pi k}. \tag{12.3}$$

🔆 **Check Your Understanding.** Show the details of computing the integrals for a_k and b_k in (12.2) and (12.3).

Since $\cos k\pi = (-1)^k$ for integers $k \geq 1$, we obtain

$$b_k = \frac{1-(-1)^k}{\pi k}.$$

Thus, the Fourier series of $S(t)$ is

$$S(t) = \frac{1}{2} + \sum_{k=1}^{\infty} \frac{1-(-1)^k}{\pi k}\sin(kt) = \frac{1}{2} + \frac{2}{\pi}\sin(t) + \frac{2}{3\pi}\sin(3t) + \frac{2}{5\pi}\sin(5t) + \cdots.$$

This decomposition reveals to us that this square wave has a fundamental frequency of $1/(2\pi) \approx 0.159$ Hz and contains all odd harmonics. Moreover, the intensity of each harmonic is inversely proportional to its frequency: $A_k = \frac{2}{\pi f_k}$ dB. The fundamental frequency should be of no surprise based on the graph of $S(t)$ since the square wave itself has period 2π and, consequently, a frequency of $1/2\pi \approx 0.159$ Hz as well.

A partial plot in the frequency domain of the fundamental frequency and first three harmonics is shown in Fig. 12.5b. If you listen to what a square wave sounds like (most synthesizers have a square wave sound built-in) then the spectrum we discovered here will make complete sense. A square wave basically sounds like a richer and buzzier sine wave, which lines up with all of the harmonics present and the more rigid shape of the time domain graph.

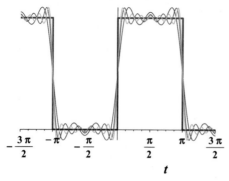

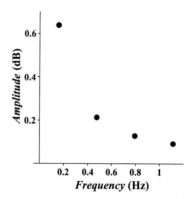

(a) Plot of the 2π−periodic square wave $S(t)$ (blue), along with partial sum approximations of the Fourier series decomposition: first 5 terms (red), first 9 terms (green), and first 13 terms (magenta).

(b) Partial plot of the spectrum (frequency domain) of the square wave $S(t)$. Points corresponding to the fundamental frequency and first three harmonics are shown.

Figure 12.5

Exercise 12.6. The function

$$S(t) = \begin{cases} t & -\pi \leq t \leq 0 \\ -t & 0 < t \leq \pi \end{cases}$$

represents a 2π−periodic wave called the **triangle wave**. Plot $S(t)$. Find the Fourier series representation of $S(t)$ and use it to create a plot in the frequency domain that includes the first five frequencies. Compare the harmonics present in the triangle wave to those present in the square wave from Example 12.5.

12.3 The Fourier transform for non-periodic signals

Recall that the Fourier series representation of a signal is only valid if the signal is periodic. This is fine to analyze sounds such as sine waves, square waves, triangle waves, etc. However, in general, periodic sounds are only a theoretical concept and in real life most sounds are either **quasi-periodic** (repeat in a noticeably similar but not exact manner) or **aperiodic** (neither periodic nor quasi-periodic). As a result, the Fourier series representation of periodic signals that we have discussed so far does not apply to most signals found in practice. However, this does not mean that the theory and intuition behind Fourier series is not useful. In fact, it is the foundation for finding a representation for non-periodic signals.

We can extend the idea of a Fourier series to non-periodic functions through a tool called the **Fourier Transform**. To see how this works we will first extend our definition of the Fourier series in equation (12.1) to a periodic function with period T (instead of just 2π). This is easily done by the following algebraic manipulations. First, for mathematical convenience, we rewrite (12.1) using Euler's formula as

$$S(t) = \sum_{k=-\infty}^{\infty} A_k e^{i2\pi kt}. \tag{12.4}$$

Exercise 12.7. Derive a formula for the complex coefficients A_k using Euler's formula and the formulas (11.2) for the coefficients a_k and b_k.

Suppose we now change the variable to t' such that $t \equiv \frac{2\pi t'}{T}$. Substituting into (12.4) yields

$$S(t') = \sum_{k=-\infty}^{\infty} A_k e^{i2\pi kt'/T}.$$

This leads us to the definition of the Fourier transform. Intuitively we can think of the Fourier transform as the limit of the Fourier series of a function whose period approaches infinity ($T \to \infty$). In particular, think of replacing the discrete coefficients A_k with continuous coefficients $\hat{S}(\xi)d\xi$ while letting $k/T \to \xi$. Then change the sum to an integral to obtain

$$S(t) = \int_{-\infty}^{\infty} \hat{S}(\xi)e^{2\pi i\xi t}d\xi. \tag{12.5}$$

Here, the function $\hat{S}(\xi)$ is called the **Fourier transform** of $S(t)$ and is defined by the formula

$$\hat{S}(\xi) = \int_{-\infty}^{\infty} S(t)e^{-2\pi i\xi t}dt. \tag{12.6}$$

This formula creates the relationship between the time-domain function S and its corresponding frequency-domain function $\hat{S}$, *regardless if the function S is periodic or not*. We will illustrate calculating the Fourier transform by hand with the following simple example.[2]

Example 12.8. Calculate the Fourier transform for the **square pulse** function given by

$$S(t) = \begin{cases} 1 & |t| \leq \frac{1}{2} \\ 0 & |t| > \frac{1}{2}. \end{cases}$$

Solution. Before we begin, note that $S(t)$ is aperiodic and so calculating a Fourier series is not possible. However, it is possible to calculate the Fourier transform. Using (12.6) we have

$$\hat{S}(\xi) = \int_{-\infty}^{\infty} S(t)e^{-2\pi i\xi t}dt$$

$$= \int_{-1/2}^{1/2} 1 \cdot e^{-2\pi i\xi t}dt$$

$$= \int_{-1/2}^{1/2} [\cos(-2\pi \xi t) + i\sin(-2\pi \xi t)]dt.$$

Using the fact that cosine is an even function and sine is odd we obtain

$$\hat{S}(\xi) = \int_{-1/2}^{1/2} \cos(2\pi \xi t)dt - i\int_{-1/2}^{1/2} \sin(2\pi \xi t)dt = \frac{\sin \pi\xi}{\pi\xi} - 0.$$

---

[2]In general, calculating the Fourier transform is a complex task that is far beyond the scope of this text and is not necessary for the goal of this chapter.

Thus,

$$\hat{S}(\xi) = \frac{\sin(\pi\xi)}{\pi\xi}.$$

The function $S(t)$ and its corresponding Fourier transform $\hat{S}(\xi)$ are shown in Fig. 12.6.

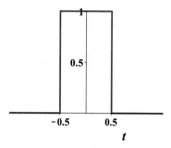

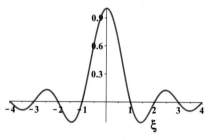

(a) Plot of the square pulse wave on $[-1/2, 1/2]$.

(b) Fourier transform of the square pulse wave.

Figure 12.6

12.4 The Discrete Fourier Transform

Physically sounds are continuous waves. Before computers, sound was recorded using analog methods that exactly replicate the continuous sound waves by storing the signal in some form of physical medium, such as a wire or tape. With the advent of computers, analog recording methods were replaced by digital methods where sound data is recorded and stored on a computer. However, due to how computers read and save information, it is not possible for a computer to process continuous data. As a result, computers record and save sound through a process called **sampling**, in which sound data measurements are captured and stored at particular discrete times. A basic example of this idea is shown in Fig. 12.7. Here, a sine wave (red curve) is sampled at 16 equally spaced points (blue solid circles). Sound sampling is an important topic in its own right in signal processing, and so we will not dive into the many intricacies here. The key point is that because digital media is saved in a discrete format, we now need a way to apply the ideas of the Fourier transform to discrete signals rather than continuous.

The mentality behind developing a discrete version of the Fourier transform is similar to that of calculating a Riemann sum instead of an integral. Given a sequence of data $(x_0, x_1, \ldots, x_{N-1})$, we define the **discrete Fourier transform (DFT)** by:

$$X_k = \sum_{n=0}^{N-1} x_n e^{-2\pi i k n/N}, \quad k = 0, 1, 2, \ldots, N-1. \tag{12.7}$$

Likewise, the inverse DFT is defined by

$$x_k = \sum_{n=0}^{N-1} X_n e^{2\pi i k n/N}, \quad k = 0, 1, 2, \ldots, N-1. \tag{12.8}$$

The rigorous derivation of the DFT is done using the idea of a so called **delta function**, which looks like a single spike centered at 0. The rigorous analysis of the Fourier transform and DFT is covered in any dedicated Fourier analysis text and is again beyond the scope of our goals here.

One important note about the DFT is that the range of k, meaning the number of detected frequencies, is limited in applications by what is called the **Nyquist Limit**. Named after electronic engineer Henry Nyquist, the Nyquist limit is defined to be half of the sampling rate of a digital signal processing system.[3] We will illustrate this in the following example.

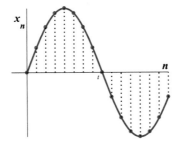

Figure 12.7. A basic example of sampling a continuous signal (red) at equally spaced times.

Example 12.9. Let $S(t) = 2\cos(2\pi t) + 3\cos(4\pi t)$. Suppose we sample $S(t)$ at a frequency of $f_s = 4$ times per second from $t = 0$ to $t = 0.75$ (see Fig. 12.8).

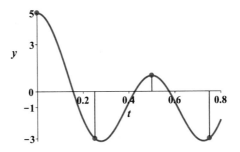

Figure 12.8. The function $S(t)$ (red) is sampled at 4 times per second on the interval $[0, 0.75]$.

Then the values of the samples are:

$$x_0 = S(0) = 5, \quad x_1 = S(.25) = -3, \quad x_2 = S(.5) = 1, \quad x_3 = S(.75) = -3.$$

Applying the DFT formula (12.7) gives us

$$X_k = \sum_{n=0}^{3} x_n e^{-2\pi i k n/4}, \quad k = 0, 1, 2, 3.$$

---

[3] In practical applications, this implies that the sampling rate should be twice the highest frequency of interest in an analog signal processing.

Now as mentioned earlier, the range of k is restricted by the Nyquist limit, which is defined to be half the sampling rate, that is the maximum k that can be detected at a given sampling rate f_s is $k_{max} = f_s/2$. In particular, for this example, the Nyquist limit is $k_{max} = f_s/2 = 4/2 = 2$ and so we really have

$$X_k = \sum_{n=0}^{3} x_n e^{-2\pi i k n/4}, \quad k = 0, 1, 2.$$

Writing out the sum for each value of k and substituting in the sample values above yields the following results:

$$X_0 = x_0 e^0 + x_1 e^0 + x_2 e^0 + x_3 e^0 = 0$$
$$X_1 = x_0 e^0 + x_1 e^{-\pi i/2} + x_2 e^{-2\pi i} + x_3 e^{-3\pi i/2} = 4$$
$$X_2 = x_0 e^0 + x_1 e^{-\pi i} + x_2 e^{-2\pi i} + x_3 e^{-3\pi i} = 12$$

Note that the k values here are unitless, but correspond to particular frequencies. To convert these values of k into frequencies measured in Hz, we need to take in account the sampling rate f_s of the original signal:

$$f = \frac{k \cdot f_s}{N}, \quad k = 0, 1, 2, \ldots, N - 1. \tag{12.9}$$

For this particular example:

$$f = \frac{k \cdot 4}{4} = k, \quad k = 0, 1, 2, 3. \tag{12.10}$$

We can now plot the frequencies using the rescaled horizontal axis measured in Hz:

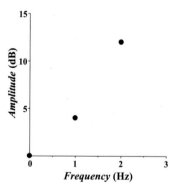

Let us now interpret and analyze the results of this example. We see from the plot of the DFT coefficients that the fundamental frequency is 1 Hz. This should be of no surprise since this can be determined by just looking at the function $S(t)$. There is also one overtone that appears at 2 Hz. The corresponding amplitudes of these two frequencies are 4 and 12. However, based on $S(t)$, we see that the frequency 1 Hz should correspond to an amplitude of 2 and that 2 Hz should correspond to an amplitude of 3. *So why do the DFT results tell us something different?* This is because the DFT scales the amplitude by a factor of N (**why?**). So, in fact, the actual magnitude of each frequency corresponding to the original signal is $|X_k|/N$:

$$1 \text{ Hz} : \frac{|X_1|}{4} = \frac{4}{4} = 1,$$

$$2 \text{ Hz} : \frac{|X_2|}{4} = \frac{12}{4} = 3.$$

This now coincides with exactly what we would expect to see based on the definition of the signal $S(t)$.

Exercise 12.10.

(a) Suppose that you are given a 24,000 sample discrete signal that is 2 minutes long. Based on the Nyquist limit, what is the highest frequency that can be detected using the DFT? How many samples would you have needed to detect a frequency of 1000 Hz? How about 5000 Hz?

(b) Write a function that inputs the duration of a signal and a frequency, and outputs the number of samples needed to capture that frequency.

12.4.1 DFT computer implementation. In the previous example we calculated by hand the DFT for a very simple situation. In practice the number of samples is often very large. For example, a common sampling rate for recording studios is 44,100 samples per second. This means that a 3 minute song is composed of 7,938,000 samples! Therefore, calculating the DFT by hand is clearly impractical. Instead, programming and numeric computation platforms are used to perform this task. Most software, such as Maple and MATLAB, include a built-in DFT function. This will be explored in the following lab.

In applications, computing the DFT from the definition is often too slow to be practical, even using modern day computers. For this reason, more advanced algorithms called "fast Fourier transforms" (FFTs) have been developed. Using a FFT manages to reduce the order of computations required from $\mathcal{O}(N^2)$ to $\mathcal{O}(N \log N)$, where N is the data size. This is quite an impressive feat that significantly reduces the time required to calculate the DFT of a signal, especially for large data sets (that is, for large values of N).

12.4.2 Lab 16: Time Domain to Frequency Domain Using the DFT/FFT. The audio files needed for this lab can be downloaded using the following link: `https://github.com/bobbydd21/bookaudio.git`.

Problem formulation. In this chapter's lab you will analyze and compare the frequencies present in a pitch played on guitar versus a piano by transforming the given audio signals from the time domain to the frequency domain. The specific CAS commands needed to perform this lab vary based on the software being used, but code samples for Maple and MATLAB are provided below.

Suggested directions.

Part 1. Use a built-in function to read in both audio files.

Directions

- In Maple, the function to read in audio files is "read". This function is part of the "AudioTools" package. Your line to read in audio files in Maple should look like "$A := Read(\text{"filename"})$". This function returns an array of the sampled data now saved to the variable A. The sampling rate, total number of samples, and duration of the audio file are printed.

- In MATLAB, the function to read in audio files is "audioread" and this line should look like "$[y,Fs]=\text{audioread('filename')}$". This function reads in data from the file named filename, and returns sampled data, y, and the sample rate for that data, Fs.

Part 2. Based on the information obtained in Part 1, calculate the Nyquist limit for each of the two files. Is the sampling rate used to record these audio files sufficient to capture all frequencies that can be detected by the human ear?

Part 3. Calculate the DFT of the sample data for each file.

Directions

- In Maple, the function "$DFT(A[..,1])$" computes the DFT of the first column of A. Note that the original audio file is in stereo (two channels) and so A is an array containing two columns (one for each channel). For simplicity, the term "$[..,1]$" is added to only consider the DFT of the first channel.

- In MATLAB, the function "$fft(y(:,1))$" computes the DFT of the first column of y. Again, we only consider the first channel of y.

Part 4. Plot the sound file in the time domain.

Directions

- In Maple, the simplest method to obtain a plot of the signal is to use the "Preview" function: "$Preview(A[..,1], output = embed)$".

- In Matlab, your code to plot the signal should look like: "$plot(t,y(:,1))$," where t is a vector containing the times in seconds that correspond to the sample data of the signal. (Hint: To define the vector t, use the sampling rate and total number of samples found in Part 1.)

Part 5. Plot the absolute value of the DFT in the frequency domain.

Directions

- In Maple, one method to plot the DFT is to use the "listplot" function from the "Plots" package. Using this function, your code to plot the DFT should look like: "$display(listplot(B[1 .. N/2]))$", where $N := length(A[..,1])$ and $B := abs(A[..,1])$. You may need to use the "view" plot option to adjust the window of your plot.

    - The issue with this plotting method is that the horizontal axis is not scaled correctly. In order to properly scale the horizontal axis, we need to take in account the sampling rate of the original signal, as we did in equation (12.9) of Example 12.9. This can easily be done in Maple through the following steps:

    (a) Define a function that computes the formula in equation (12.9):

$$\text{func} := j \mapsto \frac{44100 \cdot (j - 1)}{N}.$$

(b) Define the horizontal axis vector to be $f := Vector(N/2, func)$.

(c) Plot the DFT coefficients using the vector f as the independent variable: $plot(f, B[1..N/2])$. You may need to use the "view" plot option to adjust the window of your plot.

- In MATLAB, the easiest method to plot the DFT is through the following steps:
  (a) Let the variable $yfft = abs(fft(y(:, 1)))$ be the absolute value of the DFT of the vector y.
  (b) Discard half of the points by setting $yfft = yfft(1 : Nsamps/2)$, where $Nsamps$ is the total number of samples.
  (c) Prepare the horizontal axis by defining the vector
  $$f = Fs * (0 : Nsamps/2 - 1)/Nsamps.$$
  (d) Finally create a figure using the plot function: $plot(f, yfft)$. You may need to use the "xlim" option to adjust the window of your plot.

Part 6. For each signal, identify the fundamental frequency and first three overtones. Briefly explain and compare your results between the two signals.

💡 **Check Your Understanding.** When plotting the DFT in Part 4 of the above lab, why is only the first half of the DFT points used?

12.5 Fourier series in signal processing

We saw in the previous section that decomposing a periodic sound signal into its Fourier series representation gives information about the frequency content and energy spectrum of the sound wave. The same is true for periodic signals of a different nature, for instance, electrical signals. There are two major tasks in signal processing. The first is eliminating high-frequency noise through a process called **filtering**, or **denoising**. The second is **data compression**. Fourier series can be used as a tool for implementing these tasks. One approach to denoising is to express a signal as a Fourier series and throw away the high-frequency terms. The goal of signal compression is to transmit a signal in a way that requires minimal data transmission without losing essential information about the signal. This can be done by keeping only terms with amplitudes that are larger than some specific tolerance. In practice, these tasks are typically implemented using the DFT.

In this short section we just show graphical illustrations of these two signal processing operations.[4] Fig. 12.9 shows the effect of denoising the signal

$$f(t) = \exp(-\cos^2(t)\sin(2t) + 2\cos(4t) + 0.4\sin(t)\sin(50t))$$

when only few first terms of the trigonometric expansion are retained.

Qualitatively, the difference between denoising and compression is that in denoising the terms with frequency higher than some threshold are removed, while in compression the terms with amplitude (energy) smaller than some tolerance are removed. The amount of compression is usually defined as a certain percent of the discretized signal. For example, if a discretized signal is sampled at $2^8 = 256$ equally spaced time nodes, and 80% of the sampled values related to low amplitude components are set

---
[4]Figures 12.9 and 12.10 are produced using data from the book [3].

to zero, the assembled compressed signal has an 80% reduction. Fig 12.10 shows the effect of 80% compression on a signal.

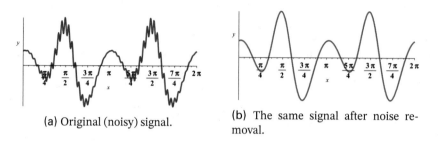

(a) Original (noisy) signal.

(b) The same signal after noise removal.

Figure 12.9. Filtering: A signal before and after implementing the denoising operation.

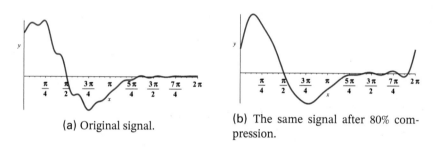

(a) Original signal.

(b) The same signal after 80% compression.

Figure 12.10. Compression: A signal before and after implementing the compression process.

12.6 Glossary

- **Pure tone** – A sinusoidal sound wave of a single frequency represented by a function of the form $S(t) = A\cos(kt - \phi)$, where A is the amplitude, ϕ is the phase shift, and $f = k/(2\pi)$ is the frequency.

- **Principle of Superposition in sound waves** - When two or more waves overlap in space, the resultant wave is mathematically the algebraic sum of the individual waves. This principle implies that all sounds can be constructed from the building blocks of pure tones.

- **Fundamental frequency** – The lowest frequency in a sum of pure tones. Frequencies greater than the fundamental frequency are called **harmonics** or **overtones**.

- **Time domain** – Used to represent the amplitude or intensity of a signal as a function of time. The **frequency domain** represents the amplitude as a function of frequency.

- **Discrete Fourier Transform (DFT)** – Given a sequence of data $(x_0, x_1, \ldots, x_{N-1})$, the DFT is defined by

$$X_k = \sum_{n=0}^{N-1} x_n e^{-2\pi i k n / N}, \quad k = 0, 1, 2, \ldots, N-1.$$

Likewise, the inverse DFT is defined by

$$x_k = \sum_{n=0}^{N-1} X_n e^{2\pi i k n / N}, \quad k = 0, 1, 2, \ldots, N-1.$$

- **Nyquist Limit** – A numerical characteristic of a sampling process equal to half of the **sampling rate**. It limits the number of frequencies that can be detected by the DFT.

Part 3

Probability and Statistics

13

Probability and Statistics Basics

Probability theory is concerned with the theoretical analysis of random phenomena. Informally, probability of a random event can be described as chances for the event to occur in some experiment. The word "experiment" is interpreted in probability theory as some process or action that can be repeated, at least hypothetically, under identical conditions and may lead to different outcomes on different trials.

Probability theory was introduced as a rigorous mathematical discipline by A.N. Kolmogorov in 1933 using an axiomatic approach.[1] This axiomatic foundation formed the basis for modern theory. Since then probability theory has been considerately developed and refined by many outstanding mathematicians and physicists, including the mathematical giant David Hilbert. More recent developments are stimulated by problems in mathematical physics and the need for probabilistic modeling in various applied fields, such as data analysis and artificial intelligence. For more on modern probability and its history see [22].

Probability theory is essential as a mathematical foundation for statistics. Statistics deals with the collection, organization, analysis, interpretation, and presentation of large sets of data related to many human activities. The concept of probability in statistics is still debated. The most common interpretations of probability are the classical, frequentist, and Bayesian.

The **classical interpretation**, going back to Laplace, applies to a random experiment with a finite set of all possible results that can be partitioned into elementary, equally likely outcomes due to symmetry or some other considerations. The probability of an event in this random experiment is defined as the ratio of the number of outcomes favorable for the event to the number of all elementary outcomes. Textbook examples of such experiments are tossing an ideal coin or rolling an ideal die.

---

[1] Kolmogorov's monograph is available in English as "Foundations of Probability Theory" (Chelsea, New York, 1950).

The **frequentist interpretation** of probability is adopted by traditional statistical methodology. It applies to experiments that can be repeated as many times as necessary (which is not always the case). The probability of an event is then approximated by the long-term frequency of the event occurrence.

The **Bayesian interpretation** is based on the idea of the degree of belief in the chance that an event will occur. It is a subjective estimate by an observer or a reasonable expectation that represents a state of knowledge about the event. Although somewhat controversial, the Bayesian interpretation underlies a powerful theory of statistical inference that is widely used in applications.

In this chapter, we will briefly review some basic terms of probability theory and recall a few important discrete and continuous probability distributions. In the last two sections we state and explore two fundamental theorems in probability theory called the Law of Large Numbers (LLN) and the Central Limit Theorem (CLT). Examples and exercises in these sections illustrate the two theorems using samples from the standard distributions introduced in this chapter. The sampling relies on computer simulations using appropriate built-in CAS functions, which at this point we will treat as "black boxes." Behind-the-scenes, these functions implement *pseudorandom number generation* algorithms (PRNGs) that emulate properties of sequences of random numbers. We will describe a very simple PRNG algorithm in the next chapter of this module and also try to make these statistical simulation techniques a little bit more transparent by taking a closer look at the underlying mathematics of these algorithms. The last chapter of the module includes a mini project and two labs that involve statistical simulation and analysis of a particular type of random process called a random walk. Hopefully, by this point, the underlying mathematics of the corresponding CAS commands will be more clear.

13.1 Review: Some basic concepts of probability

The axiomatic approach to probability theory is centered on the concept of a **probability space** that captures mathematically the idea of an experiment with random outcomes. We introduce this concept only for discrete random variables. A **random variable** (RV) is a variable that takes certain values defined by outcomes of a random experiment.

The rigorous definition of a continuous random variable and its probability space relies on the concept of a **measurable function** and is beyond the scope of this module. Intuitively, a continuous RV is one that can take on any value in an interval on the number line. For example, consider an experiment that involves twirling a spinner.[2] All possible outcomes of this experiment comprise the set of all angles that the final position of the spinner makes with some fixed direction. The range of possible angles can be chosen as the interval $[0, 2\pi)$. The interval of possible values of a continuous random variable could be infinite.

13.1.1 Discrete random variables.

Example 13.1. Consider the experiment of flipping three identical, ideal coins. Each coin lands either heads up or tails up. Thus, there are eight elementary, mutually exclusive outcomes of this experiment, which can be modeled as the set (in self-explanatory

---

[2]An idealized spinner is considered as a straight line segment having no width and pivoted at its centre.

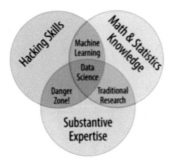

Figure 13.1. Venn diagram for data science suggested by the data scientist Drew Conway [7].

notation)
$$S = \{HHH, HTH, HHT, HTT, TTT, TTH, THT, THH\}.$$

Since we assume that the coins are ideal and identical, the probability of each elementary outcome is taken to equal 1/8. We call an **event** any subset[3] of S. For example, $A = \{HHT, HTH, THH\}$ is the event "exactly two heads" and $B = \{HHH, TTT\}$ is the event "same outcomes".

Some experiments have a countably infinite set of possible outcomes. For example, one can toss a coin until "tails" appears for the first time. The number of possible tosses is $n = 1, 2, \ldots$. Here is a formal description of a **discrete probability space**.

Definition 13.2. A discrete probability space is a triple (S, E, P), where

- S is a **sample space**, a finite or countably infinite collection $\{x_j, \ j \in I\}$ of points in some mathematical space; I being a set of indices. The points represent all elementary outcomes of a certain probabilistic experiment.

- E is a collection of all subsets of S called **events**.

- P is a function defined on E, $P : E \to [0, 1]$, such that $P(S) = 1$. For any event $A \in E$, $P(A)$ is the sum of $p_j \equiv P(x_j)$ for all $x_j \in A$. (In the case of a countably infinite set A, $P(A)$ is the sum of the series $\sum_{j=1}^{\infty} p_j$.)

📖 Relations between events in simple experiments can be illustrated using Venn diagrams. In a Venn diagram events are depicted by circles. Composite events are interpreted using operations on sets, such as union, intersection, and negation. Venn diagrams are also used to visualize relations among sets of various kinds. For instance, Fig. 13.1 shows a Venn diagram for data science suggested by a data scientist Drew Conway.

Definition 13.3 (Discrete random variable). A discrete RV is a variable that takes on one of the values in a countable set of numerical values determined by the outcomes of a random experiment.

---

[3]A set A is a subset of a set B (notation: $A \subset B$) if all elements of A are also elements of B.

For instance, a RV X related to Example 1 can be defined as the number of tails in the experiment. Possible values of the RV are integers from zero to three with probabilities

$$P(X = 0) = 1/8, \ P(X = 1) = 3/8, \ P(X = 2) = 3/8, \ P(X = 3) = 1/8.$$

💡 **Check Your Understanding.** Make lists of the elementary outcomes that comprise the events (a) $X = 1$ and (b) $X = 2$.

A list of all values of a discrete RV X and the corresponding probabilities is called a **probability table**. The same information can be represented by a so called **probability mass function** (PMF) defined by the equations $P(X = x_j) = p_j$, where x_j runs over all possible values of the random variable. A probability table or mass function completely characterizes a discrete RV. Any set of pairs of numbers (x_j, p_j), $j = 1, \ldots, n$, such that $p_j > 0$ and $\sum_{j=1}^{n} p_j = 1$ defines a valid probability table of an abstract discrete RV.

Exercise 13.4. Consider rolling a pair of ideal six-sided dice. Let the RV X be the sum of pips on the top faces of the dice.

(a) Make a probability table for the RV X.

(b) Find $P(X \leq 4)$.

There are also numerical characteristics termed **statistics** that quantify certain properties of a discrete RV X. Here are three of such characteristics:

- **Expectation (expected value, mean)**: $E(X) = \sum_{j=1}^{n} x_j p_j$ – The probability weighted average of all possible values of X.

- **Variance**: $Var(X) = E\big((X - E(X))^2\big)$ – The expectation of the squared deviation of a random variable from its mean. The formula simplifies to $Var(X) = E(X^2) - E(X)^2$.

- **Standard deviation**: $SD(X) = \sqrt{Var(X)}$.

Example 13.5. Consider rolling an ideal six-sided die. Let the RV X be the number of pips on the top face of the die. Calculate $E(X)$, $Var(X)$, and $SD(X)$.

Solution. Clearly, there are six possible values of the RV X and

$$P(X = k) = 1/6, \ k = 1, \ldots, 6.$$

We have

$$E(X) = \frac{1}{6} \sum_{k=1}^{6} k = \frac{7}{2},$$

$$Var(X) = E(X^2) - E(X)^2 = \frac{1}{6} \sum_{k=1}^{6} k^2 - \left(\frac{7}{2}\right)^2 = \frac{35}{12},$$

$$SD = \frac{\sqrt{105}}{6}.$$

⚡ *Check Your Understanding.* Show the calculation of $SD(X)$.

Some formulas for calculating statistics of sums and products of RVs assume independence of the RVs, meaning that the realization of one does not affect the probability distribution of the other. Such RVs are called **independent RVs**. Henceforth, we will be using the acronym "i.i.d." for independent, identically distributed random variables.

Properties of the expectation and variance. It can be shown that

- $E(cX) = cE(X)$.

- Expectation of the sum of RVs is equal to the sum of the expectations of these variables.

- Expected value of the product of two independent random variables is equal to the product of the expected values of these variables.

- $Var(cX) = c^2 Var(X)$.

- Variance of the sum of independent RVs is equal to the sum of the variances of these variables.

⚡ *Check Your Understanding.* Calculate $E(X)$, $Var(X)$, and $SD(X)$ for the RV X in Exercise 13.4. Use CAS assistance but do not use CAS commands that return these statistics.

13.2 Some discrete probability distributions

Discrete uniform distribution

We say that a RV has a discrete uniform distribution if the variable assumes a finite number n of values each with the same probability. Clearly, the probability of taking any value is $1/n$. A simple classical example is the RV X "Number of dots on the upper face" in rolling a fair six-sided die. This RV has a uniform distribution with $n = 6$.

Binomial distribution. A binomial RV is defined as the number of occurrences of a certain event in n independent, identical experiments (trials), with the same probability of the event in each trial. The event of interest is often termed a "success". The complement of the success, which is the set of all outcomes when the success does not occur, is called a "failure." If p is the probability of a success in one experiment, the probability of a failure is $q = 1 - p$. We write $X \sim Binom(n, p)$ if a RV X has a binomial distribution with parameters n, p. For example, consider the experiment of rolling a fair, six-sided die n times and define a RV X as the number of times when there are m dots on the upper face of the die. In this case we would write $X \sim Binom(n, 1/6)$.

Calculating the PMF of a binomial RV involves the concept of a **combination**[4] from the mathematical discipline called **enumerative combinatorics**. The number of combinations of k items from a set of n items is given by the binomial coefficient

$$\binom{n}{k} = \frac{n!}{k!\,(n-k)!}.$$

---

[4]A combination is an unordered collection of k items from a set of n items.

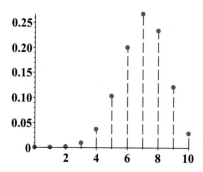

Figure 13.2. Probability mass function for the random variable $X \sim$ $Binom(10, 0.7)$.

The formula $P(X = k) = \binom{n}{k}p^k q^{n-k}$, $k = 0, \ldots, n$, defines the PMF of a binomial RV with parameters n, p.

Exercise 13.6. Prove the following identities.

(a) $\displaystyle\sum_{k=0}^{n} \binom{n}{k} = 2^n$

(b) $\displaystyle\sum_{k=0}^{n} \binom{n}{k} p^k q^{n-k} = 1$

Notice that the second formula shows that the defined PMF of a binomial RV is valid.

Example 13.7.

(a) Consider a RV $X \sim Binom(1, p)$. The PMF of this RV is defined by the equations $P(X = 1) = p, P(X = 0) = q$. This PMF defines the so called Bernoulli distribution. It is easy to show (**do this!**) that $E(X) = p$, $Var(X) = pq$.

(b) If $Y \sim Binom(n, p)$, then $Y = \sum_{k=0}^{n} X_j$, where the X_j's are i.i.d. RVs, each having Bernoulli distributions with parameter p. Therefore, by the properties of the expectation and variance, $E(Y) = np$, $Var(Y) = npq$.

Exercise 13.8.

(a) Use CAS assistance to construct a probability table of the RV $X \sim Binom(7, 0.6)$.

(b) Make a CAS function **my_binom(n, p)** that takes an integer n and a number $p \in (0, 1)$ and returns

- the probability table $\{(k, P(X = k))\}_{k=1}^{n}$, for $X \sim Binom(n, p)$
- $E(X), Var(X)$, and $SD(X)$
- a plot of the PMF of X

Test your function on the RV X in part (a). Your output plot will be similar to the one in Fig. 13.2.

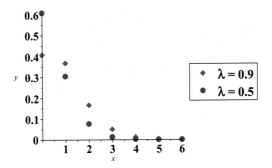

Figure 13.3. Probability mass functions of two Poisson distributions.

The Poisson distribution. A discrete RV X is said to have a Poisson distribution with parameter $\lambda > 0$ if its PMF is defined as

$$P(X = k) = \exp(-\lambda)\frac{\lambda^k}{k!}, \ \ k \in \mathbb{N}^0.$$

In this case we write $X \sim Poisson(\lambda)$. It is an example of a variable with a countably infinite set of values.

Exercise 13.9. Consider a RV $X \sim Poisson(\lambda)$.

(a) Show that all probability values of X add up to one.

(b) Show that $E(X) = \lambda$.

Hint: Use the power series for the exponential function to complete the following derivation of $E(X)$:

$$E(X) = \sum_{j=0}^{\infty} jP(X = j) = \exp(-\lambda) \sum_{j=0}^{\infty} j \cdot \frac{\lambda^j}{j!} = \lambda \exp(-\lambda) \sum_{j=1}^{\infty} \frac{\lambda^{j-1}}{(j-1)!} = \ldots.$$

(c) Show that $Var(X) = \lambda$.

Hint: Complete the following derivation

$$Var(X) = E(X^2) - (E(X))^2 = \exp(-\lambda) \sum_{j=0}^{\infty} j^2 \frac{\lambda^j}{j!} - \lambda^2 = \ldots.$$

The Poisson probability model applies to a RV that counts repeated independent occurrences of a certain event in a fixed interval of time or a certain region of space, provided that the average rate of the event occurrences is constant. A classic textbook example of a Poisson distribution that fits a real historical data set is described in the next example.

Example 13.10. During World War II the city of London was divided into 576 small areas of one-quarter square kilometer each and the number of areas hit by flying bombs exactly k times was recorded. The only observed number of hits to these small areas were $0, 1, 2, 3, 4, 5$. The table below shows the numbers of areas m_k hit exactly k times for $k = 0, \ldots, 5$.

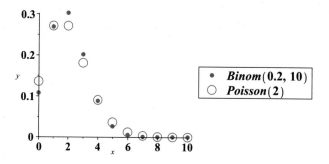

Figure 13.4. Illustration of the relationship between $Binom(n, p_n)$ and $Poisson(\lambda)$ distributions for large n and fixed $\lambda = n \cdot p_n$.

k, **number of hits**	0	1	2	3	4	5
m_k, **number of areas**	229	211	93	35	7	1

There were a total of 537 hits, so the average number of hits per area was 0.9323. Let a RV X be the number of hits of a small area. The observed frequency of the event $X = k$ is $m_k/576$. We assume that $X \sim Poisson(0.9323)$. The table below shows the numbers of hits $537 \cdot P(X = k)$ predicted by this Poisson model. The theoretical predictions are remarkably close to the observations. (This example uses the frequentist interpretation of probability.)

k, **number of hits**	0	1	2	3	4	5
m_k, **number of areas**	226.7	211.4	98.5	30.6	7.1	1.6

Remark 13.11. Notice that for $k > 5$, the predicted number of areas rounds to zero. For example, the predicted number of areas hit six times is $537 \exp(-0.9323) \cdot 0.9323^6/6! \approx 0.193 \approx 0$.

Other examples of random variables that fit a Poisson distribution include

- The number of pieces of mail from a wide range of sources per interval of time (day, hour, month) arriving independently of one another.

- The number of decay events per second from a radioactive isotope, provided the half-life of the substance is in thousands of years.

Relation between binomial and Poisson distributions. It can be proved that if a sequence $\{np_n\}_{n=1}^{\infty}$, $p_n \in (0, 1)$, converges to a finite number λ as $n \to \infty$, then for large values of n the distribution of a RV $X \sim Binom(n, p_n)$ can be approximated by the Poisson law with parameter λ. To prove this statement, one should show that $\lim_{n\to\infty} P(X = k)$ is equal to $\exp(-\lambda)\lambda^k/k$, the probability that a RV $Y \sim Poisson(\lambda)$ takes value k. We believe the proof technique is within reader's reach. (**Try it!**) The goal of the next exercise is to illustrate the relation between binomial distributions and an appropriate Poisson distribution when the condition stated above is met.

Exercise 13.12.

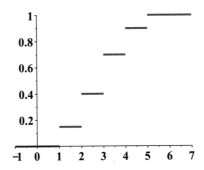

Figure 13.5. Plot of the cumulative distribution function for a discrete random variable.

(1) Make a CAS function **binom_to_poisson(n, p)** that takes parameters n, p and computes the PMF of the RV $X \sim Binom(n, p)$ and the PMF of the RV $Y \sim Poisson(np)$. The function returns a figure with plots of the PMFs of the two RVs.

(2) Experiment with the function made in part (a). Make figures for various pairs of parameters n, p_n with increasing values of n and fixed products np_n. Show one of the figures. Your figure will be similar to Fig. 13.4.

13.3 About continuous probability distributions

Continuous random variables are usually measurements. Examples include height, weight, the amount of sugar in a certain kind of fruit, or the time between the arrival of customers to a service facility.

A probability distribution of a continuous RV is completely defined by one of two related nonnegative functions – a **cumulative distribution function** (CDF) or a **probability density function** (PDF). In this section we will work with RVs with continuous or piecewise continuous PDFs. A CDF of any RV X is defined by the equation $F(x) = P(X \leq x)$. For a continuous RV with CDF F, the corresponding probability density function is the derivative of F.

The probability that a RV X with PDF f takes values in an interval (a, b) can be found as

$$P(a < X < b) = \int_a^b f(x)d x. \tag{13.1}$$

It is assumed in this formula that one or both limits of integration can be infinite.

☼ **Check Your Understanding.** Complete the following formulas for a continuous RV X with PDF f. Explain.

- $CDF(x) \equiv F(x) = \ldots$ (The formula should involve the PDF f.)

- $P(X = c) = \ldots$

- $\int_{-\infty}^{\infty} f(x)d x = \ldots$

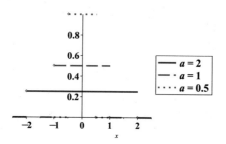

Figure 13.6. Probability density functions for three uniform RVs defined on the intervals $[-a, a]$.

Summary statistics for a continuous RV are defined in a similar way to that of a discrete RV, but the sums are replaced with integrals:

$$E(X) = \int_{-\infty}^{\infty} x f(x) d\,x, \tag{13.2}$$

$$Var(X) = \int_{-\infty}^{\infty} x^2 f(x) d\,x - E^2(X). \tag{13.3}$$

⬚ **Check Your Understanding.** Use the properties of expectation to derive the formula (13.3) from the definition of the variance as $E\big((X - E(X))^2\big)$.

Exercise 13.13. Make a function *discrete_rv_cdf*(L) that takes a list L of pairs of real numbers (x_k, p_k) such that

- $x_k \neq x_j$, for $k \neq j$

- $p_k \in (0, 1)$ and $\sum_{k=1}^{n} p_k = 1$

and returns a plot of the CDF of the discrete RV defined by the PMF $P(X = k) = p_k$. Include verification that the given p_k values sum to one. Test your code on a concrete list of data. The output of your function will be similar to the plot in Fig. 13.5.

⬚ **Check Your Understanding.**

(a) What are the values of a discrete RV with CDF depicted in Fig. 13.5?

(b) Is a CDF of any discrete RV continuous from the right or from the left at the dicontinuity points?

Uniform distribution. We say that a continuous RV X has a uniform distribution on the interval $[a, b]$ and write $X \sim U(a, b)$ if its PDF is defined as $f(x) = 1/(b - a)$ for $x \in [a, b]$ and zero otherwise.

Exercise 13.14.

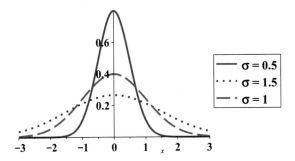

Figure 13.7. Normal curves for PDFs with zero expectation and various standard deviations.

(a) Show that if $X \sim U(a, b)$ and $[c, d] \subset [a, b]$ then $P(X \in [c, d])$ depends only on the length of the subinterval $[c, d]$.

(b) Calculate the expected value, the variance, and the standard deviation of the continuous RV $Y \sim U(-a, a)$.

(c) Fig. 13.6 shows three plots of the PDFs of the RVs $Y_a \sim U(-a, a)$ with $a = 2, 1, 0.5$. Show by **symbolic** calculation the effect of halving the parameter a on $Var(Y_a)$. Write a complete sentence describing the change in the variance of Y_a when the parameter a is replaced with $a/2$.

Normal distribution. A normal distribution is arguably the most important probability distribution. A RV X with normal distribution has PDF of the form

$$f(x) = \frac{1}{\sigma\sqrt{2\pi}} \exp\left(-\frac{(x-a)^2}{2\sigma^2}\right),$$

and we write $X \sim N(a, \sigma)$. It can be shown that $E(X) = a$ and $SD(X) = \sigma$. The distribution of a RV X with $a = 0$ and $\sigma = 1$ is called **standard normal** and the RV X the **standard normal RV**. Fig. 13.7 shows the effect of the standard deviation on the shape of the PDF graph of a normal RV. A PDF graph of any normal RV is called a **normal curve**.

In the next exercise you will make functions that solve two basic problems for a standard normal RV: evaluating its CDF at a given input and inverting the CDF given its value.

Exercise 13.15. Consider the standard normal RV $X \sim N(0, 1)$.

(a) Make a CAS function **normal_cdf(c)** that takes a real number c and returns the probability $P(X \leq c)$, that is, $CDF(c)$. Test your function using an example with a known answer.

(b) Make a CAS function **inverse_normal_cdf(p)** that takes a number $p \in (0, 1)$ and returns the number c such that $P(X < c) = p$. Test your function using an example with a known answer.

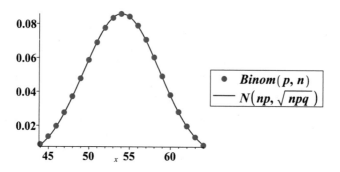

Figure 13.8. Approximation of the binomial distribution by the normal distribution with the same mean and variance.

Remark 13.16. The function in part (b) of Exercise 8 is an example of a so called **quantile function**, or inverse cumulative distribution function. If the CDF F of a continuous RV X is strictly increasing, then the equation $P(X \le x_p) = p$, $0 < p < 1$, has the unique solution $x_p = F^{-1}(p)$ called the pth **quantile**. The map $p \mapsto x_p$ is the quantile function. When p is given as a percent, the term **percentile** is used for x_p.

Relation between binomial and normal distributions. Consider a binomial RV $X \sim Binom(n, p)$ and a normal RV Y $Y \sim N(np, \sqrt{npq})$ with the same mean and variance as X, that. It is known that for a large number of trials n the distribution of X can be approximated by the distribution of Y. The next example illustrates this relation.

Example 13.17. Consider a RV $X \sim Binom(90, 0.6)$. Since $E(X) = np = 54$ and $Var(X) = npq = 21.6$, we will use the PDF of the normal RV $Y \sim N(54, \sqrt{21.6})$ for approximating the PMF of X. Fig. 13.8 shows that that the normal distribution fits the binomial distribution like a glove. We have chosen the range $44 \le k \le 64$ for the values of X. The motivation for this choice is the **68-95-99.7 rule** for the normal distribution. One of the claims of this rule is that $P(|Y - E(Y)| < 2\sigma) \approx 0.95$. In this example, $E(Y) = 54$, $2\sigma \approx 9.3$. Therefore the interval $[44, 64]$ used in Fig. 13.8 covers values of the RV Y that occur slightly more than 95% of the time.

Exercise 13.18. Write a function *binom_to_normal*(**n**, **p**) that takes parameters n, p of a RV $X \sim Binom(n, p)$ and returns a figure with the normal curve defined by parameters $a = np$, $\sigma = \sqrt{npq}$ in the range $a \pm 2\sigma$ and the PMF of X. Test your function for a binomial distribution with a sufficiently large n value to obtain a figure similar to Fig. 13.8.

13.4 Law of Large Numbers

The Law of Large Numbers (LLN) is part of a group of theorems sometimes called the Law of Averages. It relates the sample average of a large number of values of a RV X and the expected value of X. To develop some intuition and make this informal statement more precise, let us consider a simple example.

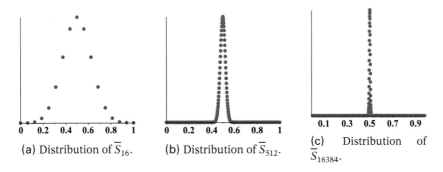

(a) Distribution of $\overline{S}_{16}$.　(b) Distribution of $\overline{S}_{512}$.　(c)　Distribution　of $\overline{S}_{16384}$.

Figure 13.9. Probability mass functions for the RVs $\overline{S}_n$ in Example 13.19.

Example 13.19. Let X_j, $j = 1, \ldots, n$, be i.i.d.[5] Bernoulli RVs with probability of success $p = 0.5$. Consider a new RV $\overline{S}_n = \sum_{j=1}^{n} X_j/n$. Calculate the expectation and the standard deviation of this random variable. Find and plot the distribution of the RV $\overline{S}_n$ for $n = 16, 512, 16384$.

Solution. To solve the problem we will use our knowledge about the RV S_n.

(a) We can show that $\overline{S}_n$ has the same expectation as any of the X_j, that is, $E(\overline{S}_n) = p$. (**Do this!**) Notice that the RV $S_n = \sum_{j=1}^{n} X_j$ has a binomial distribution with parameters p, n, and $Var(S_n) = npq$. It follows that $Var(\overline{S}_n) = pq/n$. (Show this!)

(b) It is easy to find the PMF of the RV $\overline{S}_n$ since $P(\overline{S}_n = \frac{k}{n}) = P(S_n = k)$. Thus the PMF of $\overline{S}_n$ is $k/n \mapsto \binom{n}{k} p^k (1 - p)^{n-k}$.

(c) Fig. 13.9 shows three plots of the PMF of the RV $\overline{S}_n$ for the given values of n. The figure suggests that as $n \to \infty$ there is some kind of convergence of the theoretical distributions of $\overline{S}_n$ to the **constant distribution**.[6] This constant is the common mean value of the RVs X_j, which is 0.5 for this example. This observation is supported by our calculation of the standard deviation of the RV $\overline{S}_n$, which goes to zero as $n \to \infty$.

An intuitive meaning of the LLN theorem is that the average of values of i.i.d. RVs observed in a large number of trials is expected to become close to the common expected value of these RVs, and will tend to become closer as more trials are performed.

Exercise 13.20. Consider the experiment of rolling a fair six-sided die n times. Let Y_j, $j = 1, \ldots, n$, be the RVs "number of times when 3 pips are on the upper face of the die in the jth trial." Introduce a RV $\overline{S}_n = \sum_{j=1}^{n} Y_j/n$. Find and plot the distribution of the RV $\overline{S}_n$ for $n = 20, 200, 2000$.

In formal terms, the LLN theorem can be stated as follows:

---

[5]Recall that this acronym stands for "independent, identically distributed"
[6]A constant distribution is a distribution of a "variable" that takes a specific value with probability one.

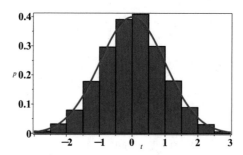

Figure 13.10. The standard normal curve superimposed on the histogram of 1000 samples of the random variable S_{500}^* in Example 13.24.

Theorem 13.21 (Law of Large Numbers). *Let X_j, $j = 0, 1, 2, \ldots$, be i.i.d. RVs with finite expected value μ and finite variance. Let a RV $\overline{S}_n$ be the average of the X_j, $j = 0, 1, 2, \ldots, n$. Then for any $\varepsilon > 0$*

$$P(|\overline{S}_n - \mu| \geq \varepsilon) \to 0$$

as $n \to \infty$.

The proof of the theorem relies on Chebyshev's inequality, which is a remarkable statement in its own right.

Theorem 13.22 (Chebyshev's inequality). *Let X be a RV with finite expected value μ and finite variance $Var(X) = \sigma^2$. For any $c > 0$*

$$P(|X - \mu| \geq c) \leq \frac{\sigma^2}{c^2}.$$

Qualitative meaning of Chebyshev's inequality: The larger the constant c is, the less is the probability that the deviation of the RV X from its mean is greater than c.

13.5 Central Limit Theorem

The **Central Limit Theorem** (CLT) is the fundamental theorem in probability theory and the cornerstone of statistics. It implies that some probabilistic methods and tools that work for normal distributions can be applied towards approximating solutions of problems involving other types of distributions. There are many versions of the CLT. Intuitively, the theorem states that, under some fairly mild assumptions, the distribution of sample means approximates a normal distribution as the sample size gets larger, regardless of the population's distribution[7] from which the samples are drawn. Alternatively, the theorem claims that the **standardized sum** of a large number of i.i.d. RVs is approximately the standard normal RV. The standardization of a random variable X is defined as

$$X^* = \frac{X - E(X)}{SD(X)}.$$

Here is a more formal version of the CLT.

---

[7]In statistics, the term **population** refers to the total set of observations or measurements of interest that can be made.

Theorem 13.23 (Central Limit Theorem). *Let X_j, $j = 1, \ldots, n$, be i.i.d. RVs with $E(X_j) = a$, $SD(X_j) = \sigma$. Let $S_n = \sum_{j=1}^{n} X_j$. Then the distribution of the standardized RV S_n^* for sufficiently large n is approximately standard normal, **regardless of the underlying distribution of the RVs X_j.***

The amazing part of this theorem is shown in bold. The next example demonstrates the CLT when the underlying distribution of RVs X_j is known.

Example 13.24. Consider again the experiment of rolling a fair, six-sided die. Suppose the experiment is repeated n times. Let X_k denote the RV "number of pips on the upper face of the die in the kth experiment". Let $S_n = \sum_{k=1}^{n} X_k$. Do the following.

(a) Find the statistics $E(S_n)$ and $SD(S_n)$.

(b) Use CAS assistance to generate 1000 standardized values of the RV S_{500}^*.

(c) Plot a histogram of S_{500}^* and the standard normal curve in one figure.

Solution. We solve the problem in three steps.

(a) As was shown in Example 13.5, each X_k is a discrete uniformly distributed RV defined by the PMF $P(X_k = j) = 1/6$, $j = 1, \ldots, 6$, with $E(X_k) = 3.5$ and $SD(X_k) = \sqrt{105}/6$. By the properties of the expected value and variance, $E(S_n) = n\,E(X_k)$ and $SD(S_n) = SD(X_k)/\sqrt{n}$.

(b) Using a CAS command for sampling from a discrete uniform distribution, we made a function that takes a positive integer n, generates n random samples from the discrete uniform distribution with six possible values, and returns the corresponding value of the standardized RV S_n^*. Using this function, the sample of 1000 values of S_{500}^* was obtained.

(c) Fig. 13.10 shows the histogram of these values and the superimposed standard normal curve.

The two mini projects below demonstrate the CLT for a binomial distribution and a continuous uniform distribution.

13.5.1 Two mini projects illustrating CLT.

Mini project: Approximating standardized sums of i.i.d. binomial RVs. Consider a RV $X \sim Binom(n, p)$. Follow the steps in the previous example to demonstrate the Central Limit Theorem. Specifically,

Step 1. Make a CAS function **standardized_sum(n, p, N)** that generates N values of the RV X and returns a corresponding value of the standardized sum S_N^* of these values.

Step 2. Make a CAS function **clt_for_binom(n, p, N, M)** that uses the function made in Step 1 to obtain M values of S_N^* and returns a figure with the histogram of these values and a superimposed standard normal curve. Test your code with parameters $n = 20$, $p = 0.7$. Experiment with the values of N and M to produce a figure similar to Fig. 13.10.

Mini project: Approximating standardized sums of i.i.d. continuous uniform RVs. Consider a RV $X \sim U(a, b)$. Follow the steps in the previous mini project to demonstrate the Central Limit Theorem for the standardized sums of the values of X.

13.6 Glossary

- **Discrete random variable** (RV) – A variable that takes on one of the values in a countable set of numerical values determined by the outcomes of a random experiment. A discrete RV can be defined by its **probability mass function** (PMF), which is a set of equations $P(X = x_j) = p_j$, where x_j runs over all possible values of the random variable and all the p_j values sum to one.

- **Independent RVs** X **and** Y:

$$P(X = x, Y = y) = P(X = x) \cdot P(Y = y) \quad \text{(discrete case)}$$
$$P(X < x, Y < y) = P(X < x) \cdot P(Y < y) \quad \text{(continuous case)}$$

- **Cumulative distribution function** (CDF) of a RV X – A function defined by the equation $F(x) = P(X \leq x)$. For a continuous RV, the corresponding **probability density function** (PDF) is the derivative of F.

- **Law of Large Numbers** (LLN) – Intuitively, the law states that the average of values of independent, identically distributed (i.i.d.) RVs observed in a large number of identical trials is expected to become close to the common expected value of these RVs and will tend to become closer as more trials are performed.

- **Central Limit Theorem** (CLT) – Intuitively, the theorem claims that under certain fairly general assumptions, the standardized sum of a large number of i.i.d. RVs is approximately standard normal.

14

Computer Simulation of Statistical Sampling

Computer simulation is used to analyze and predict behaviors of systems in many fields when physical simulation is too complicated, too expensive, or simply impossible. It relies on a mathematical model of the system of interest to imitate the system behavior. Moreover, computer simulation allows one to check the reliability of the chosen mathematical models for situations where it is not possible to obtain related data by observing or measuring the quantities of interest. This is why computer simulations have become a useful tool for mathematical modeling of many real-life processes.

Typically, computer simulations are computationally intensive, but the availability of powerful and user-friendly modern mathematical software makes it possible to simulate and analyze some moderately complex models as a laboratory exercise. In this chapter we will design and perform computer simulations of simple probabilistic models. There are two labs in this chapter. The first shows the CLT in action on a real data set with an unknown distribution and the second is on the classic Buffon's Needle Problem.

In some exercises in the previous chapter, CAS commands for sampling from different probability distributions have been used as "black boxes". In this chapter we will take a closer look at the underlying mathematics behind the computer implementation of random sampling.

14.1 Random number generation

Devices called hardware random number generators are used to produce truly random numbers through some chaotic physical processes, such as quantum phenomena or fluctuations of some atmospheric parameters. However, most random numbers used in computer programs are pseudorandom, which means that they are generated in a predictable, deterministic fashion using mathematical formulas. Several algorithms for **pseudorandom number generators** (PRNGs) of integers have been developed over the years. To get just a glimpse of these methods, we will take a brief look at what

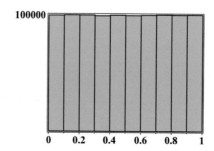

Figure 14.1. Histogram of 10^6 pseudo-random floats in $(0,1)$ obtained using the multiplicative congruential algorithm.

is called a **linear congruential generator** (LCG). LCGs were considered standard in the second half of the 20th century.[1] An LCG starts with an initial value x_0, called the **seed**, and then recursively computes successive values x_n using a certain linear recurrence. A multiplicative LCG uses the recurrence $x_{n+1} = a\, x_n$ mod m with some integer parameters a, m; another LCG uses the recurrence $x_{n+1} = a\, x_n + b$ mod m. The parameter m is called the **modulus**.

To obtain a list of pseudorandom floats from $U(0,1)$, integers in a list generated by a PRNG are divided by the modulus m. But how do we obtain a list of random values from a general interval (a, b)? For any two finite intervals, there is a linear function that maps one interval to the other. This fact will be constructively used in the next exercise.

Exercise 14.1. Make a CAS function ***random_floats*(n, a, b)** that takes an integer n and generates a sequence of n random numbers from $U(0, 1)$ using an appropriate CAS command. The function should then map this sequence into the interval (a, b) and return the result.

The next example demonstrates that a "homemade" PRNG constructed without a deeper understanding of randomness can be very bad.

Example 14.2. Using CAS assistance, we applied the multiplicative LCG with $a = 512$ and modulus $m = 203$ to generate a sequence of 30 integers starting with the seed $x_0 = 8388607$:

$$(171, 59, 164, 129, 73, 24, 108, 80, 157, 199, 185, 122, 143, 136,$$
$$3, 115, 10, 45, 101, 150, 66, 94, 17, 178, 192, 52, 31, 38, 171, 59).$$

As expected, our homemade PRNG with arbitrary chosen parameters is very bad. After just 28 numbers the generator starts to repeat the sequence. However, the example demonstrates the general idea of how the multiplicative congruential algorithm works. In principle, with a clever choice of parameters, the maximum number of different integers that the algorithm can produce is $m - 1$, one less than the modulus.

---

[1]More recent algorithms, like the **Mersenne Twister**, avoid some of the problems with earlier deterministic PRNGs.

Using the multiplicative congruential algorithm with $a = 7^5$, $x_0 = 1410999733$, and $m = 2^{31} - 1$, we made a list L of 10^6 random integers. Fig. 14.1 shows the histogram of corresponding floats from the interval $(0, 1)$. We applied a CAS command that detects repeated elements in our list L. The command returned the empty list, that is, all 10^6 integers in L are different.

Remark 14.3. There are algorithms for generating random integers from a bounded interval $[a, b)$, $a \in \mathbb{N}^0$, $b \in \mathbb{N}$. The simplest case is when $b - a = 2^n$. Then we can generate a random integer r from uniformly distributed integers in the interval $[a, b)$ using the following algorithm:

(1) Toss a coin n times writing 0 for tails and 1 for heads to obtain a binary number. (For large n one can use computer assistance in this step.)

(2) Convert the obtained binary number to the decimal number k. (Note that the random decimal integers k vary from 0 to $2^n - 1$.)

(3) Set $r = a + k$.

Exercise 14.4. Let X be an unbiased Bernoulli RV with values $x \in \{0, 1\}$. Make a CAS function that generates N triples (m_1, m_2, m_3) of X-values, makes a list of the corresponding decimal numbers using the formula $n = 2^2 m_1 + 2 m_2 + m_3$, and returns a histogram of these numbers. What is the range of possible values of n in this exercise?

📖 Although all existing PRNGs fail the test to be truly random, special generators are carefully designed for particular applications, such as cryptography, to meet certain security requirements. Due to the numerous uses of random numbers in both the sciences and real-life applications, the search for high-quality PRNGs algorithms and their implementation is a very active area in computer science.

14.2 Lab 17: CLT and LLN in action: Life expectancy in the world population

The term **population** refers to the total set of observations or measurements of interest that can be made. A characteristic of interest is considered as a random variable. Often, neither the probability distribution of this RV nor the statistics of this distribution are known. However, it is possible to construct a so called **sampling distribution** of the statistic of interest and obtain some information about its value for the total population. The CLT and LLN play a crucial role in implementing this task.

In the lab below you will be dealing with a real data set containing the life expectancy at birth for 202 world countries. This set of numbers is our population of interest. If we draw a large number of samples of the same size n from this population and calculate the means (averages) of the samples, we obtain the **sampling distribution of means**. The CLT tells us that for a large n the sampling distribution of averages is approximately normal. Then the LLN comes into play and assures that the mean of the sampling distribution (the mean of averages) approaches the population mean as n grows.

Remark 14.5. The CLT does not answer the question of how large the sample size n needs to be for the normal approximation to be valid. The answer depends on the population distribution of the sample data (which is often unknown). A general rule of thumb is that we can be confident of the normal approximation whenever the sample size n is at least 30.

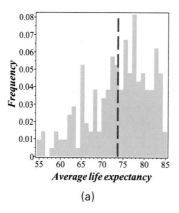

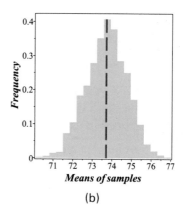

(a)   (b)

Figure 14.2. (a) The population distribution and (b) a sampling distribution for Lab 16. The dashed lines depict the mean values of the distributions.

14.2.1 Lab 17: Life expectancy in the world population. Consider the list P of life expectancy at birth for 202 world countries (see Appendix A). This data defines the distribution of the average life expectancy predicted for 2022. This is the population used in the lab.

Problem formulation and suggested directions.

Part 1. Calculate $\mu = E(P)$ and $\sigma = SD(P)$. Plot the line $x = \mu$ superimposed on the histogram of the data P.

Directions: We suggest to use 30 bins for the histogram. Your figure will be similar to Fig. 14.2a. Notice that the population distribution does not resemble any standard textbook distributions.

Part 2. Make a function *sampling_distr*($\mathbf{P}, \mathbf{n}, \mathbf{M}$) that takes a list of data P of length N, natural numbers $n < N$, and M, and does the following:

- Generates M samples of length n from P and makes a list S of averages of these samples. This is a sampling distribution of averages (means).
- Makes a histogram of the sampling distribution.
- Computes the mean m and the **standard error**[2] s of the sampling distribution of averages.

Directions: To obtain n random values from P, use an appropriate CAS function to generate n random integers in the range $[1, M]$, where M is the length of P. Then use these values as indices to extract corresponding elements of P.

---

[2]This is the term used for the standard deviation of a sampling distribution.

Part 3. Run the function made in the previous part with $n = 50$ and $N = 1000$. Superimpose the line $x = m$ on the histogram of the data S. Your figure will be similar to Fig. 14.2b.

Part 4. Run the function made in Part 2 with parameters $N = 1000$ and n taking values $n_1 < n_2 < ... < n_{10}$, $n_1 \geq 50$. Then analyze the trends in the corresponding sequences m_j, s_j, $j = 1, 2, 3, 4$. Compare the values m_j with μ and the values s_j with $\sigma/\sqrt{n_j}$, and comment on what you see.

14.3 Sampling from non-uniform distributions (optional)

When sampling from a specific probability distribution other than a uniform one, computer generated uniformly distributed floats are transformed in certain ways depending on the distribution of interest. This section describes a transformation for this task called the **distribution function technique**.

Functions of random variables. Suppose X is a RV with PDF $f(x)$ and u is a certain function defined on the range of X. We want to find the PDF $g(y)$ of the RV $Y = u(X)$. The next example illustrates the "mechanics" of the method. Along the way, we will discover conditions on the function u needed to make this problem solvable.

Example 14.6. Let $f(x) = 4x^3$ for $x \in [0, 1]$ and $f(x) = 0$ otherwise. Let X be a RV with PDF $f(x)$. Find the PDF $g(y)$ of the RV $Y = X^3$.

Solution. Since $f(x) \geq 0$ on $[0, 1]$ and $\int_0^1 f(x)dx = 1$, the function f is a valid PDF. The corresponding CDF is

$$F(x) = \begin{cases} 0 & \text{if } x \leq 0 \\ x^4 & \text{if } 0 < x \leq 1 \,. \\ 1 & \text{if } x > 1 \end{cases}$$

We will solve the problem in two steps. First, we will find the CDF $G(y)$ of the RV Y. Then, differentiating the function G, we will obtain the PDF g.

(1) By definition,
$$G(y) = P(Y \leq y) = P(X^3 \leq y). \tag{14.1}$$
The function $u : x \mapsto x^3$ is strictly increasing on $[0, 1]$ and its inverse $v : y \mapsto \sqrt[3]{y}$ is also strictly increasing. Therefore $X^3 \leq y$ if and only if $v(X^3) \leq v(y)$. Applying the function v to both sides of the inequality $X^3 \leq y$ in (14.1), we find
$$G(y) = P(X^3 \leq y) = P(X \leq \sqrt[3]{y}).$$
But the right-hand side of this equation is just the CDF of the RV X evaluated at $\sqrt[3]{y}$! Thus $G(y) = F(\sqrt[3]{y}) = \sqrt[3]{y^4}$.

(2) Differentiating G, we find the solution to the problem: the PDF of the RV $Y = X^3$ is $g(y) = 4\sqrt[3]{y}/3$.

In general, let $F(x)$ be a CDF of a RV X and $Y = u(X)$. The following conditions on the function u ensure that the RV Y has a PDF in the class of piecewise continuous functions:

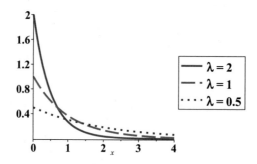

Figure 14.3. Three probability density functions of the exponential distributions.

- The function u is defined on the sample space (all possible values) of the RV X and is strictly increasing.

- The function v, the inverse of u, is piecewise differentiable on the sample space of Y.

Under these conditions, you can find the PDF of the RV Y as follows:

(1) $G(y) = P(Y \leq y) = P(u(X) \leq y) = P(X \leq v(y)) = F(v(y))$

(2) $g(y) = \frac{dG}{dy}$

Here is another application of the distribution function technique.

Example 14.7. Consider a RV $X \sim U(0, 1)$ and the function $u(x) = -\ln(1 - x)/\lambda$, with $\lambda > 0$. Find the PDF of the RV $Y = u(X)$.

Solution.

(1) Check the conditions for applying the distribution function technique:

- Function u is defined for $x \in (0, 1)$ and is strictly increasing (**show this**).
- The inverse of the function u is $v(y) = 1 - \exp(-\lambda y)$ (**show this**). The function v is (infinitely) differentiable for all $y \in \mathbb{R}$.

(2) $G(y) = P(Y \leq y) = P(-\ln(1 - X)/\lambda \leq y) = P(X \leq 1 - \exp(-\lambda y)) = 1 - \exp(-\lambda y)$.

(3) $g(y) = \frac{dG}{dy} = \lambda \exp(-\lambda y)$.

Problem solved. The RV Y with PDF g found in this example has a so called **exponential distribution**.

📓 *Check Your Understanding.* Explain why $P(X \leq 1 - \exp(-\lambda y)) = 1 - \exp(-\lambda y)$ in this example.

Let us take a closer look at this continuous probability distribution.

Exercise 14.8. Consider an exponential RV X defined by the PDF

$$g(x) = \lambda \exp(-\lambda x).$$

Use CAS assistance to do the following.

(a) Show that the function g is a valid PDF.

(b) Find the statistics $E(Y)$ and $SD(X)$.

(c) Make a figure with three PDFs from the exponential family with parameter values of your choice. Your figure will be similar to Fig. 14.3.

Example 14.7 shows that if X is a uniform RV with sample space $(0, 1)$, then the RV $Y = -\ln(1-X)/\lambda$ has an exponential distribution with parameter λ. An algorithm implementing the distribution function technique takes a sample x from the uniform distribution $U(0, 1)$ and returns the value of the function $x \mapsto -\ln(1 - x)/\lambda$, which is just the inverse of the CDF $F(x) = 1 - \exp(-\lambda x)$ for an exponential distribution with parameter λ. (This is the quantile function for the exponential distribution.) Fig. 14.4 shows the histogram of 10000 values calculated according to this procedure with $\lambda = 2$ and the graph of the PDF $f(x) = \lambda \exp(-\lambda x)$ for the corresponding exponential distribution.

☞ **Check Your Understanding.** Show that $x \mapsto -\ln(1-x)/\lambda$ is the quantile function for the exponential distribution with parameter λ.

Exercise 14.9. Solve the problem stated in Example 14.7 but with $\lambda = 1.5$. Make a figure similar to Fig 14.4. Use a CAS command to sample from the uniform distribution $U(0, 1)$ and then apply the inverse transformation method. You may use interactive coding or make a function that takes a parameter λ and an integer n, and returns a figure with a histogram of n random values from the exponential distribution with parameter λ and the graph of the PDF of the exponential distribution with the same parameter.

Exercise 14.10. Consider a RV X defined by the piecewise PDF function $f(x) = 3x^2$ for $0 \leq x \leq 1$ and zero otherwise. Find the PDF of the RV $Y = X^2$.

By this point it should be clearer what exactly is going on under-the-hood when using CAS commands to generate random samples from specific statistical distributions. From now on these commands should seem a bit more transparent than a complete "black box".

14.4 Monte Carlo methods for finding integrals and areas

14.4.1 Finding integrals of univariate functions. Let g be a continuous function on an interval $[a, b]$, and suppose we want to compute the integral $J = \int_a^b g(x)\, dx$, $a < b$. The Fundamental Theorem of Calculus provides an exact formula for finding definite integrals if the integrand has an elementary antiderivative. However, there are elementary functions that do not have such antiderivatives. A textbook example is the function $\exp(-x^2)$.

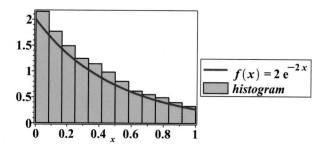

Figure 14.4. Exponential PDF with parameter $\lambda = 2$ and a histogram of 10^5 samples from this distribution calculated using the inverse transformation method.

There are various deterministic numerical integration methods for approximating definite integrals. This section describes a surprising and fascinating probabilistic method for numerical integration using random numbers. The method, called **Monte Carlo (MC) integration**, relies on the computer simulation of random numbers. There are versions of MC integration that use various techniques of simulation. In this chapter, only uniform sampling will be used.

📖 MC integration is just one of a broad class of Monte Carlo algorithms that rely on random sampling. The method is particularly useful for approximating **multidimensional integrals**.

MC integration: Intuition

Recall that the average value of a finite collection of numbers c_j, $j = 1, \ldots, N$, is $\sum_{j=1}^{N} c_j/N$, and the average value of an integrable function g on an interval $[a, b]$ is defined as

$$g_{ave} = \frac{1}{b-a} \int_a^b g(x)\, d x. \tag{14.2}$$

Consider the following algorithm:

- Generate a large sample $\{x_j\}_{j=1}^{N}$ drawn from $U(a, b)$.

- Calculate the values $g_j \equiv g(x_j)$, $j = 1, \ldots, N$.

- Solve the equation $g_{ave} \approx \sum_{j=1}^{N} g_j/N$ for the integral, where g_{ave} is defined by equation (14.2), to obtain the approximation formula

$$\int_a^b g(x)d x \approx \frac{b-a}{N} \sum_{j=1}^{N} g_j. \tag{14.3}$$

MC integration: A more accurate probabilistic explanation. To give a more rigorous justification of formula (14.3) we employ the following two facts.

Fact 1. Let X be a RV defined by a continuous or piecewise continuous PDF f and $Y = g(X)$, where g is a continuous function defined on the sample space of X. Then

$E(Y) = \int_{-\infty}^{\infty} g(x)f(x)d\,x$. In particular, if $X \sim U(a, b)$, then

$$E(g(X)) = \frac{1}{b-a} \int_a^b g(x)d\,x. \tag{14.4}$$

☐ ☀ **Check Your Understanding.** Use formula $E(Y) = \int_{-\infty}^{\infty} g(x)f(x)d\,x$ to justify (14.4).

Fact 2. Consider a large sample $S = \{g_j\}_{j=1}^N$ with $g_j \equiv g(x_j)$ and x_j drawn from the uniform distribution $U(a, b)$. The **sample mean** $\bar{g} = \sum_{j=1}^N g_j/N$ gives an estimate of $E(g(X))$. The estimate tends to become better when the size of the sample increases. (**This is just the LLN in action!**)

Approximating $E(g(X))$ in equation (14.4) with the sample mean $\bar{g}$ and solving for the integral, we retrieve formula (14.3). Thus, this more formal argument supports our intuitive approach to the problem.

Example 14.11. Use the MC integration method to estimate the integral

$$J = \int_0^1 \exp(-x^2)d\,x.$$

Solution. The exact value of the integral is not available since the integrand does not have an elementary antiderivative. A high-precision numerical approximation of the integral gives $J \approx 0.7468241328$. Using CAS assistance, three samples $S_j, j = 1, 2, 3$, of different sizes from $U(0, 1)$ have been generated. The estimates of the given integral according to formula (14.3) with $g(x) = \exp(-x^2)$ are:

- For sample S_1 of size 1000: $J \approx 0.75546$.

- For sample S_2 of size 10000: $J \approx 0.74716$.

- For sample S_3 of size 100000: $J \approx 0.74662$.

Remark 14.12. Although in this particular example our approximations are getting better for samples of larger sizes, there is no guarantee that this is always so. The convergence of the sample means to the expected value of the RV $Y = g(X)$ is more subtle. We do not discuss this type of convergence here.

Exercise 14.13. Use CAS assistance to implement the MC integration method to approximate the following integrals. Notice that the limits of integration are different from the interval $(0, 1)$. Find the relative errors of the approximations using the exact or numeric values of the integrals.

(a) $\int_1^3 x \ln^2(x)d\,x$. Exact value: $2 + 9\ln(3)(\ln(3) - 1)/2$.

(b) $\int_{-1}^2 (2 + x^3)^{1/2}d\,x$. High accuracy approximation: 5.1765.

(c) (optional) $\int_0^\infty \exp(-x^2)d\,x$. Exact value: $\sqrt{\pi}/2$.

   Hint: Use the change of variable $x = \tan(t)$ to transform the infinite interval of integration into a finite one. Then apply the MC integration method.

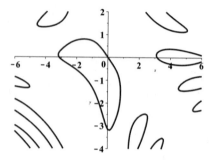

Figure 14.5. A fragment of the implicitly defined curve $\sin(x+y) - \cos(xy) - 1 = 0$.

📖 The error estimate of MC integration involves the **sample variance** s_N $\equiv \sum_{j=1}^{N}(g_j - \bar{g})^2$. It can be shown that the error of the MC estimate is bounded from above by $(b-a)s_N/\sqrt{N}$ if the sample variance remains bounded as $N \to \infty$.

Using the MC method presented in this subsection, one can find the area between the x-axis and the graph of an explicit nonnegative function. A slightly more involved MC method for finding areas bounded by a simple closed curve defined by an implicit function is introduced in the next subsection.

14.4.2 Finding area of a region bounded by an implicit curve. In some cases an implicit curve can be parametrized, that is, defined by parametric equations $x = x(t)$, $y = y(t)$, $t \in [t_1, t_2]$. If the point $(x(t), y(t))$ traverses the curve counterclockwise when the parameter t varies from t_1 to t_2, then the area of the region enclosed by the curve can be found using the formula

$$A = \frac{1}{2} \int_{t_1}^{t_2} x(t)y'(t) - y(t)x'(t)\,d\,t.$$

Sometimes it is possible to transform an implicit function into the explicit form in polar coordinates. If the region of interest is bounded by two curves defined by equations $r = r_1(\theta)$, $r = r_2(\theta)$ with $r_1(\theta) \le r_2(\theta)$, $\theta_1 \le \theta \le \theta_2$, the area of the region can be found using the formula

$$A = \frac{1}{2} \int_{\theta_1}^{\theta_2} \left(r_2(\theta)^2 - r_1(\theta)^2\right)d\,\theta.$$

However, equations of implicit curves can be too complex to transform into explicit forms. Fig. 14.5 shows part of a very complex level curve $\cos(xy) - \sin(x+y) - 1 = 0$ of the function $f(x, y) = \cos(xy) - \sin(x+y) - 1$. The curve has infinitely many components. In Example 14.15 we will find the area enclosed by one of the components of this curve. Finding the area enclosed by a complicated, implicit simple closed curve looks like a very difficult task, but random numbers come to the rescue. Here are the steps of the MC method for estimating area in such complicated cases.

(1) Enclose the region of interest into a rectangle $R = \{(x,y)|a \le x \le b, c \le y \le d\}$ with sides parallel to the coordinate axes.

(2) Generate a large number N of random points (x, y) in the rectangle with x, y being values of the uniform random variables $X \sim U(a, b)$, $Y \sim U(c, d)$, respectively.

(3) Count the number m of points (x, y) that hit the region inside the curve.

(4) The expression $\frac{m}{N} \cdot Area(R)$ gives an approximation of the area inside the curve.

About a joint probability distribution of two variables. A plane region is a two-dimensional object. To better understand why the probabilistic method for estimating areas of planar regions works, we need to first look at the simplest version of a **joint probability distribution**. For our modest purposes in this chapter, it is sufficient to describe this concept for a pair of independent uniform variables $X \sim U(a, b)$, $Y \sim U(c, d)$. Recall that $[a, b]$ and $[c, d]$ are the sample spaces of these RVs. The rectangle $R = \{(x, y) \mid a \leq x \leq b, c \leq y \leq d\}$ is the **joint sample space** of the **bivariate** RV (X, Y). This is an example of the so called **Cartesian product of two sets**, in this case, two intervals. The probability density function of the **bivariate** uniform distribution with sample space R is the constant $1/Area(R)$ for $(x, y) \in R$ and zero otherwise.[3] An event is a subset of the joint sample space. We consider here only events represented by subsets of R that are enclosed by simple closed curves and use the following fact: **The probability of an event is the ratio of the area of the subset representing the event to the area of the joint sample space.** Therefore, an estimate of the probability of an event yields an estimate of the area of the region of interest.

The question now is how do we determine whether a random point is inside the region representing an event? A simple closed curve $f(x, y) = 0$ defined by a function f that is continuous on the plane $\mathbb{R}^2$ divides the plane into two regions called the interior region and the exterior region. (It is probably hard to believe, but this statement that looks so obvious is in fact a theorem called **Jordan's theorem** whose proof by elementary considerations is rather tricky.) The function has different signs for points in the interior and exterior regions. To determine the sign of the function in the interior region, one can choose any interior point and calculate the function value at this point. Values of f at all interior points have the same sign as the value at the chosen point. The next simple problem illustrates these considerations.

Example 14.14. Two friends, Bobby and Greg, who take the metro to their jobs from the same station arrive at the station randomly between 6:00 and 6:20 am. They are willing to wait for one another for 5 minutes, after which they take a train whether they are together or alone. What is the probability of them meeting at the station?
Solution. Let the uniform RVs X, Y, each with the sample space $[0, 20]$, denote random arrival times of the two friends. The sample space of the bivariate uniform RV (X, Y) is a square with side of length 20. Fig. 14.6 shows the sample space and the event $M = $ "Bobby and Fred meet" (the green diagonal strip).

⌨ *Check Your Understanding.* Show that $P(M) = 7/16$, or about 44%.

Example 14.15. Consider one of the contours from Fig. 14.5 depicted in Fig. 14.7. The contour is defined by the implicit equation $\sin(x+y) - \cos(xy) - 1 = 0$ and, as Fig. 14.7

---

[3]This is similar to the PDF of a RV $X \sim U(a, b)$, which is a constant $1/(b - a)$ for $x \in (a, b)$ and zero otherwise.

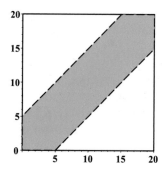

Figure 14.6. The sample space of a pair of independent identical uniform RVs and a green region representing the event of interest in Example 14.14.

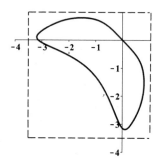

Figure 14.7. Region bounded by a simple closed curve.

shows, can be enclosed in the rectangle $R = \{(x, y) \mid -3.3 < x < 1, -3.3 < y < 1\}$. Approximate the area of the region inside the contour using the MC method.

Solution. To solve the problem, we made a CAS function **mc_area(f, M, a, b, c, d)**. The input parameters of the function are:

- function f that implicitly defines the contour by the equation $f(x, y) = 0$

- number M of random points to be generated

- parameters a, b, c, d that define a rectangle enclosing the area of interest

The function implements Steps 2 through 4 of the description of the MC method given above. We ran the function with $f(x, y) = \sin(x + y) - \cos(xy) - 1$, parameters a, b, c, d defined by the rectangle R, and $M = 10^k$, $k = 4, 5, 6$.
Results:

$$mc\_area(fcn, 10^4, -3.3, 1. - 3.3, 1) = 6.941,$$

$$mc\_area(fcn, 10^5, -3.3, 1. - 3.3, 1) = 7.096,$$

$$mc\_area(fcn, 10^6, -3.3, 1. - 3.3, 1) = 7.075.$$

A high accuracy approximation of the area found using tools of multivariate calculus and numerical integration is 7.084.

In the next exercise, the MC method will be used to approximate the number π.

Exercise 14.16. Consider a disc of radius one centered at the origin. You know that the area of the disc is π. Do the following:

(1) Enclose the disc in a square S centered at the origin with sides of length two. Make the CAS function **mc_area(g, M, a, b, c, d)** described in Example 14.15 and use it for $N = 10^k$, $k = 2, 4, 6$, and appropriate values of a, b, c, d that define S to approximate π.

(2) Choose a different square enclosing the disc and use the function made in part (a) to approximate π for the same sample sizes. Compare your results with the estimates of π in part (a) and comment on your observations.

14.4.3 Mini project: Finding areas bounded by Cassini ovals. A Cassini oval is the set of all points on the plane such that the products of the distances of any point in the set from two given points is constant. Choose as the x-axis the line passing through the two given points. Then define the y-axis to be the line passing through the midpoint of the interval between the two points. The equation of a Cassini oval is

$$\left((x + a)^2 + y^2\right)\left((x - a)^2 + y^2\right) = b^4.$$

Here a is half of the distance between the two given points. Due to dimensionality considerations, the constant product of distances is denoted by b^4. (This notation implies that a and b are in the units of length.) Simplifying, one obtains the equation

$$(x^2 + y^2)^2 - 2a^2(x^2 - y^2) - (b^4 - a^4) = 0. \tag{14.5}$$

🔆 **Check Your Understanding.** Derive equation (14.5).

Problem formulation.

Step 1. Transform (14.5) into an explicit equation in polar coordinates.

Step 2. Make polar plots of the Cassini curves for

(a) $a = 1.0, b = 1.1$.

(b) $a = 1.0, b = \sqrt{2}$.

(c) $a = 1.0, b = 2.0$.

Step 3. Consider the explicit polar equation of the Cassini curve found in Step 2 (c). Find the area enclosed by the curve using a high accuracy CAS numerical integration command. It is known that for $b > a$ the area is defined by the integral $A = 2b^2 \int_0^{\pi/2} \sqrt{1 - e^{-4} \sin^2(t)}\, d\, t$, $e = b/a$.

Step 4. Apply the MC method to estimate the area bound by the Cassini oval with $a = 1.0, b = 2.0$, and find the relative error of your approximation using the result in Step 3 as "exact".

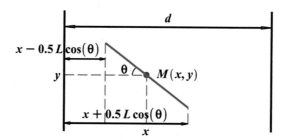

Figure 14.8. Schematic of the geometry of a random position of the needle between two parallel lines spaced by a distance d.

14.5 Lab 18: Buffon's needle problem

This is a classic problem that has an analytic solution based on the relation between the probability of a certain event and areas of the relevant plane figures. Computer simulation will be used to estimate the known theoretical probability of this event. This lab includes modeling, analysis, and algorithm construction and implementation for this classic problem.

Problem description. Let a needle of length L be thrown at random onto a horizontal plane ruled with parallel straight lines a distance d apart with $d \geq L$. What is the probability that the needle will intersect one of these lines? A schematic of a randomly placed needle is depicted in Fig. 14.8.

Mathematical model.

Notation.

- x – distance from the center of the needle to the nearest line

- d – distance between parallel lines

- L – length of the needle

- θ – orientation angle of the needle

- p – the probability of the event H = "The needle hits one of the lines"

Assumptions.

(1) Distance x is a value of the RV $X \sim U(0, d)$.

(2) Angle θ is a value of the RV $\Theta \sim U(-\pi/2, \pi/2)$.

(3) The random variables X and Θ are independent.

Analysis. The joint sample space of the bivariate RV (Θ, X) is a rectangle $S = \{(\theta, x) \mid \theta \in [-\pi/2, \pi/2], x \in [0, d]\}$ in the (θ, x)-plane. The event H occurs if and only if

$$x - 0.5L \cos(\theta) < 0 \ \text{ or } \ x + 0.5L \cos(\theta) > d. \tag{14.6}$$

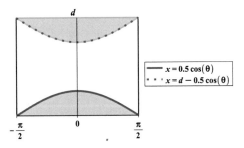

Figure 14.9. The rectangle S representing the joint sample space of the uniform RVs Θ, X and two subsets of equal area A depicting the event "the needle hits a line".

☀️ **Check Your Understanding.** Explain why the formulas in (14.6) represent the event H.

In Fig. 14.9 the conditions (14.6) define the union of two regions in S of equal area A representing the event H: below the blue curve $x = 0.5L\cos(\theta)$ and above the dotted curve $x = d - 0.5L\cos(\theta)$. The theoretical probability of the event H can be found as the ratio of the area representing the event to the area of S:

$$P(H) = \frac{2A}{Area(S)}. \tag{14.7}$$

Problem formulation.

Part 1. Find the exact expression of $P(H)$ in terms of the parameters L and d.

Part 2. Make a CAS function ***my_buffon*(N, L, d)** that takes the number of needles N of length L and the distance d between parallel lines on the plane with $d \leq L$. The function should:

- Simulate N uniformly distributed random points (x, θ) in the rectangle S and count the number of hits of the set representing the event H.
- Return the theoretical probability (14.7) that a needle hits a line and its estimate as the relative frequency of hits.

Part 3. Choose some specific values of the parameters d, L with $d < L$ and experiment with the sample size N to obtain an approximation of π with relative error less than 5%.[4]

14.6 Glossary

- **Linear congruential generator** (LCG) of random integers – An algorithm that starts with an initial value x_0 called the **seed**, and then recursively computes successive values x_n using the recurrence $x_{n+1} = a x_n \pmod{m}$ with some integer parameters a, m (multiplicative LCG), or the recurrence $x_{n+1} = a x_n + b \pmod{m}$. The parameter m is called the **modulus**.

---

[4]Notice that π is "hidden" in the denominator of the formula (14.7). It will appear when Part 1 is done.

- **Sample mean** $\overline{x}$ – The average of the given N sample values of a RV X. The **sample variance** is defined by the formula $s = \sum_{j=1}^{N}(x_j - \overline{x})^2$.

- **Joint sample space** of two continuous independent uniform RVs $X \sim U(a, b)$ and $Y \sim U(c, d)$ – The Cartesian product $[a, b] \times [c, d] = \{(x, y) \mid x \in [a, b], y \in [c, d]\}$ of the sample spaces of X and Y. An event involving the two RVs is a subset of the joint sample space. The probability of an event is the ratio of the area of the event to the area of the joint sample space (rectangle).

15

Simple Random Walks

The Law of Large Numbers and the Central Limit Theorem state certain fundamental facts about the distribution of large sums of i.i.d. random variables. In this chapter, we will consider certain **sequences of sums** of i.i.d. RVs called **random walks**. Random walks are key examples of so called random or **stochastic processes**. They have been used to model a variety of different phenomena in physics, chemistry, biology and beyond.

Definition 15.1. Let $\{X_j\}_{j=1}^N$ be independent RVs, each having a Bernoulli distribution with the sample space $\{-1, 1\}$. A **simple one-dimensional random walk on the integers** (1D RW) is a sequence of the RVs

$$S_n = \sum_{j=1}^n X_j, \; S_0 = 0. \tag{15.1}$$

A simple 1D RW is also called a random walk on the **integer lattice**. A similar structure on the plane, called the **square lattice**, is the two-dimensional version of the integer lattice, denoted as $\mathbb{Z}^2$. It is a regular grid with sites at all points with integer coordinates.

Definition 15.2. Let $Z_j = (X_j, Y_j)$, $j = 1, \ldots N$, be a sequence of ordered pairs of independent RVs X_j, Y_j each having a Bernoulli distribution with the sample space $\{-1, 1\}$. A **two-dimensional simple random walk on the square lattice** (2D RW) is a sequence of RVs

$$S_n = \sum_{j=1}^n Z_j, \; S_0 = 0, \tag{15.2}$$

where the RVs Z_j are added component-wise.

Any concrete sequence of values of a 1D RW or 2D RW is interpreted as a **trajectory**, or path, of a random walker (particle) moving along a line or in the plane. At each discrete moment of time, $t_0 < t_1 < \cdots < t_N$, the walker makes a move of unit

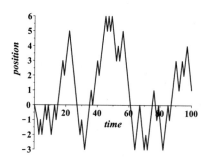

Figure 15.1. A trajectory of a simple 1D random walk with 100 steps.

length in one of a number of possible directions. The number of directions depends on the dimension of the space where the process takes place. The RV S_n is interpreted as the position of the walker on the number line or on the plane after n time moments.

If the Bernoulli distribution of the random variables X_j, Y_j in the definition of a simple random walk is unbiased, which means that the parameter p of the distribution equals 1/2, then the probability for the walker to move in any possible direction is the same. Such random walks are called **symmetric**. In this chapter we will be dealing with simple symmetric random walks unless otherwise stated.

After a brief review of some basic mathematical facts about simple 1D and 2D RW, some of these facts will be explored in exercises and two labs using computer simulation. The goal of the lab "Gambler's ruin" is to analyze a classic application of a type of random walk called a 1D RW with boundaries. The goal of the lab "Drunken sailor" is to estimate the mean distance from the initial position of a walker on the plane after n steps. Although the problem formulation in this lab looks flippant, the model is a working tool for defining an important physical parameter called the **diffusion constant**.

The presentation of some theoretical concepts and facts on random walks in this chapter is inspired by the first two lectures in the book [18].

15.1 Simple random walks on integers

Suppose the walker moves along a line. For convenience, we will depict the line as vertical and use the horizontal x-axis to show the step enumeration of the walker. The walker starts at the origin and at every integer time moment j, $j = 1, 2, \ldots, N$, flips a fair coin and moves a unit step up if the coin comes up heads or a unit step down if it comes up tails. To formally describe this process we use unbiased Bernoulli RVs X_j, $j \geq 1$, taking one of the values ± 1. The RV X_j represents the outcome at the jth moment. The position of the walker at time n is defined by equation (15.1). It is easy to show that $E(X_j) = 0$, $Var(X_j) = 1$. Fig. 15.1 shows an example of a 100-step **trajectory (path)** of this 1D simple random walk.

☼ *Check Your Understanding.* Toss a coin five times and manually plot the corresponding trajectory of the 1D random walk.

Example 15.3 (Number of trajectories of a 1D RW). For $n = 1$ there are only two possible paths with $S_1 = \pm 1$. For $n = 2$ there are four possible paths defined by the sequences

- $(1, 1)$ with $S_2 = 2$,

- $(-1, -1)$ with $S_2 = -2$,

- $(1, -1)$, $(-1, 1)$ with $S_2 = 0$.

Since the four paths are equally likely, the probability of each path is 2^{-2}. In general, for a simple symmetric 1D RW with N steps, there are 2^N paths each having probability 2^{-N}.

In the next exercise the probability distribution for the RV S_3 will be constructed by "brute force". Later in this section, a general, more clever approach to finding the probability distribution of S_n is presented.

Exercise 15.4. Complete the probability table of the RV S_3.

path	$[1, 1, 1]$	...	...	...	...	...	...	$[-1, -1, -1]$
S_3	3	...	...	...	...	...	...	-3

Various probabilistic properties of random walks can be of interest in both theory and applications. In this section we will explore only three common questions:

- On the average, how far does the walker go? Formally, what is the expected value $E(|S_n|)$?

- What is the distribution of the RV S_n?

- Does the walker always return to the starting point? If the answer is "yes", what is the probability of visiting the origin as a function of n?

15.1.1 Mini project: Mean distance of the walker from the origin.

About the expected value $E(|S_n|)$. It is easy to find the mean distance $E(|S_n|)$ of a walker from the origin for small n. For example, if $n = 2$, then using Example 15.3 we find the PMF of the RV $|S_2|$ to be $P(|S_2| = 0) = P(|S_2| = 2) = 0.5$. Thus

$$E(|S_2|) = 0 \cdot 0.5 + 2 \cdot 0.5 = 1.$$

Exercise 15.5.

(a) Use the result of Exercise 15.4 to construct the PMF of the RV $|S_3|$.

(b) Calculate the average distance $E(|S_3|)$ of the walker from the origin.

Finding $E(|S_n|)$ for large values of n is not so straightforward, but finding the **mean squared displacement** $E(S_n^2)$ is within our reach. The example below shows details of this calculation and provides some insight on the expected distance traveled by the walker. The example uses the independence of the RVs X_j, some basic algebra, and properties of the expectation.

Example 15.6. Calculate $E(S_n^2)$.

Solution.

$$E(S_n^2) = E\left(\left(\sum_{j=1}^{n} X_j\right)^2\right)$$

$$= E\left(\sum_{j=1}^{n} X_j^2 + 2\sum_{i \neq j} X_i X_j\right)$$

$$= E\left(\sum_{j=1}^{n} X_j^2\right) + 2E\left(\sum_{i \neq j} X_i X_j\right)$$

$$= n + 0 = n.$$

Since $E(S_n^2) = n$, intuitively $E(|S_n|)$ should be **of order** $\sqrt{n}$. In fact, it is known that $E(|S_n|) = c_n\sqrt{n}$ with constants $c_n \to \sqrt{2/\pi}$ as $n \to \infty$.

Exercise 15.7. Use the solution to Exercise 15.4 to find c_3.

In the mini project below the reader will estimate the average distance from the origin for a simple 1D RW using computer simulation and find an approximation of the theoretical value of the constant $c = \lim_{n \to \infty} c_n$.

Problem formulation.

Step 1. Make a CAS function *rw1_distance*(**n**) that takes an integer n, simulates and plots a trajectory $0, S_1, S_2, \ldots, S_n$ of the simple 1D RW, and returns $|S_n|$. Run your function for a particular input. Your plot will be similar to the one in Fig. 15.1.

Step 2. Make a CAS function *rw1_mean_distance*(**n, m**) that makes a list of m values of the distances produced by repeated runs of the function *rw1_distance*(**n**) and returns the average of these m values.

Step 3. Use the function *rw1_mean_distance*(**n, m**) with $m = 10000$ to make lists L_{odd} and L_{even} of estimates of the constants c_n using the values $n = k^2, k = 3\ldots, 20$. Here the lists L_{odd} and L_{even} correspond to odd and even values of n, respectively.

Step 4. Theoretical estimates for the first three terms in the lists L_{odd} and L_{even} are 0.866, 0.839, 0.827, and 0.750, 0.765, 0.773, respectively. Notice that

- L_{odd} is a decreasing sequence.
- L_{even} are an increasing sequence.
- $L_{even}[k] < L_{odd}[k]$, $k = 1, 2, 3$.

The computations done in Step 3 for large n will support these observations. It is known that the two sequences have a common limit $\sqrt{2/\pi}$. Find the best approximation of this theoretical limit based on your computations in Step 3. Round your approximation to three decimal places.

15.1.2 Distributions of the random variables S_n. In the sample space of the RV S_n there are 2^n possible sequences of +1's and -1's, and the probability of each is 2^{-n}. If the number of steps is even, $n = 2k$, the walker arrives at an even point. Formally we say that only the points $x = 2j$, $|j| = 0, \ldots, k$, are **reachable**. If the number of steps is odd, the walker arrives at an odd point. To arrive at the point $2j$, $0 \le |j| \le k$, after $n = 2k$ steps, the walker needs $k + j$ moves up and $k - j$ moves down (**why?**). There are $\binom{2k}{k+j}$ paths for this to occur. Thus

$$P(S_{2k} = 2j) = 2^{-2k}\binom{2k}{k+j}. \qquad (15.3)$$

In the next exercise, use a similar argument to construct the distribution of S_n for an odd n and combine the results for even and odd number of steps in a single formula.

Exercise 15.8.

(a) Find the probability of a walker's arrival to the point $2j + 1$, $|j| = 0, \ldots, k$, after $n = 2k + 1$ steps.

(b) Write the formula for $P(S_n = k)$ as a function of n and k given by a single expression independent on the parity of the number of steps n.

The result of part (b) of this exercise for $n = 3$ can be summarized in a table:

k	-3	-2	-1	0	1	2	3
$2^0 P(S_0 = k)$	0	0	0	1	0	0	0
$2^1 P(S_1 = k)$	0	0	1	0	1	0	0
$2^2 P(S_2 = k)$	0	1	0	2	0	1	0
$2^3 P(S_3 = k)$	1	0	3	0	3	0	1

This table shows the relation between the probability distributions of the RVs S_n and Pascal's triangle. In fact, it is simply Pascal's triangle padded with intervening zeros. This provides a basis for an iterative algorithmic construction of the probability distributions of S_n for both even and odd n. The algorithm will be implemented in the next exercise.

Exercise 15.9.

(a) Consider the table above with distributions of S_n, $n = 1, 2, 3$. Use the iterative procedure based on the relation of this table with Pascal's triangle to make a CAS function **pmf_S(n)** that takes an integer n and returns a dictionary with keys k, $0 \le |k| \le n$, and the associated values $P(S_n = k)$ of the PMF of the RV S_n.

(b) Make a CAS function **pmf_S_plot(n)** that takes a number n, calls the function **pmf_S(n)** and plots the PMF of S_n. The output plot will be similar to one of the plots in Fig. 15.2.

15.1.3 Probability of visiting the origin. Using equation (15.3) with $j = 0$, we will find the exact expression for the probability of visiting the origin after any even number of steps $n = 2k$:

$$P(S_{2k=0}) = 2^{-2k}\binom{2k}{k}. \qquad (15.4)$$

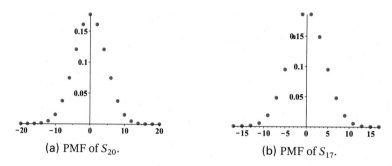

(a) PMF of S_{20}.   (b) PMF of S_{17}.

Figure 15.2. Probability distributions of S_n for an even and odd number of steps n.

This is a nice, exact expression but it is not easy to estimate how this probability changes with k just by looking at it. To find out how big or small the theoretical probability of the walker's returns to the origin is for large n, we can use Stirling's formula[1]:

$$n! \sim (n/e)^n\sqrt{2\pi n}.$$

This will be done in the next exercise.

Exercise 15.10. Use Stirling's formula and CAS assistance to show that

$$P(S_{2k} = 0) \sim \frac{1}{\sqrt{k\pi}}.$$

We can also estimate this probability using simulation.

Exercise 15.11. Make a CAS function ***rw1_returns*(k, M)** that takes positive integers k and M, simulates M values of the RV S_{2k}, counts the number of events $S_{2k} = 0$, and returns the relative frequency of the walker's visits to the origin after $2k$ steps. Choose large values of k and M and compare the theoretical probability of the event and its relative frequency found from your simulation.

Probability that the walker returns to the origin (optional)
Now we consider a simple RW with an unlimited number of steps, that is, an infinite sequence of RVs $S_n, n = 0, 1, 2, \dots$. Our goal is to find the probability of the event "The walker returns to the origin."

Lemma 15.12. *Let V be the number of returns of the walker to the origin. Then $E(V) = \infty$.*

Proof. For each $k > 0$ we define the RVs Y_{2k} that take a value of 1 if $S_{2k} = 0$ and 0 otherwise. Then $V = \sum_{k=1}^{\infty} Y_{2k}$. Notice that since

$$E(Y_{2k}) = 0 \cdot P(S_{2k} \neq 0) + 1 \cdot P(S_{2k} = 0) = P(S_{2k} = 0),$$

we have $E(V) = \sum_{k=1}^{\infty} P(S_{2k} = 0)$. It follows from Exercise 15.10 that

$$\lim_{k \to \infty} \frac{P(S_{2k} = 0)}{1/\sqrt{k}} = \sqrt{\pi}.$$

---

[1] Here the sign "$\sim$" means that the ratio of the quantities on the left and right of this sign tends to one as $n \to \infty$.

(**Show this!**) Since the series $\sum_{k=1}^{\infty} 1/\sqrt{k}$ diverges, by the Ratio Test the series for $E(V)$ diverges as well. □

Now using the lemma, we will prove the following theorem.

Theorem 15.13. *For a simple 1D RW, the probability that the walker returns to the origin infinitely often is one.*

Proof. Let q be the probability that the walker ever returns to the origin after time $n = 0$. We will show that $q = 1$ by assuming $q < 1$ and deriving a contradiction using the distribution of the RV V.

The probability that the walker never returns to the origin is $1 - q$, that is $P(V = 0) = 1 - q$. Then

$$P(V = 1) = q(1 - q)$$
$$P(V = 2) = q^2(1 - q)$$
$$P(V = k) = q^k(1 - q)$$
$$\vdots$$

It follows that

$$E(V) = \sum_{1}^{\infty} kP(V = k)$$

$$= \sum_{1}^{\infty} k \cdot q^k(1 - q)$$

$$= q(1 - q) \sum_{1}^{\infty} k \cdot q^{k-1}$$

$$= q(1 - q) \left(\sum_{1}^{\infty} q^k \right)'$$

$$= q(1 - q) \left(\frac{1}{1 - q} \right)'$$

$$= \frac{q}{1 - q} < \infty.$$

But we proved in lemma 15.12 that $E(V) = \infty$. This contradiction completes the proof of the theorem. □

15.2 Lab 19: The gambler's ruin problem

The goal of the lab is to use simulation to explore a certain game of chance. Both symmetric and non-symmetric RWs will be used in modeling this problem. In addition, the RWs start with a nonzero position and have a new feature called an **absorbing state**.

Definition 15.14. If the walk ends when a certain value of S_n is reached, then that value is called an **absorbing boundary** or **absorbing state**.

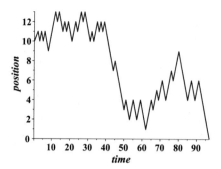

Figure 15.3. A gambler starting with $10 with the goal to double the initial capital goes bankrupt after 118 rounds.

There are different versions of the gambler's ruin problem. Here is one of the most common scenarios. A gambler enters a casino with n dollars in cash and starts playing a game where he wins with probability p and loses with probability $q = 1 - p$. The gambler plays the game repeatedly, betting $1 in each round. He leaves the game if his total fortune reaches N dollars or he runs out of money, whichever happens first. The two most typical questions are:

- What is the probability of winning?

- What is the expected number of steps until stopping?

Mathematical model. The problem can be modeled as a random walk with initial position $S_0 = n$ and two **absorbing states**, $S_t = 0$ (gambler's ruin) and $S_t = N$ (gambler's win). Notice that the time index t, the number of rounds until termination of the game, is not specified in advance.

Let w, l denote the events "win" with $p(w) = p$ and "loss" with $p(l) = 1 - p$ in one round of the game. Let W_m be the event "winning the game provided that in some round the gambler has m dollars." Note that the event W_0 is impossible, that is, $P(W_0) = 0$, and the event W_N is certain, that is $P(W_N) = 1$. For $0 < k < N$ we can write

$$W_k = w \cdot W_{k+1} + l \cdot W_{k-1}.$$

(**Think why.**) Using probability rules for the sum and product of events, along with the given probabilities of winning or losing in one round, we derive the formula

$$P(W_k) = p \cdot P(W_{k+1}) + q \cdot P(W_{k-1}). \tag{15.5}$$

💡 **Check Your Understanding.** Specialize the formula for $k = N - 1$ and $k = 1$ assuming that $p = 1/2$ (unbiased RW).

With notation $s_k \equiv P(W_k)$ and $r \equiv q/p$ we can rewrite (15.5) as the **recurrence relation**

$$s_{k+1} = (1/p) \cdot s_k - r \cdot s_{k-1}, \tag{15.6}$$

with boundary conditions $s_0 = 0$, $s_N = 1$. For $p = q$, the solution for this recurrence relation will be derived in the next exercise.

Exercise 15.15.

(a) For a simple RW, the recurrence (15.6) can be rewritten as

$$s_{k+1} = 2 \cdot s_k - s_{k-1}.$$

Use the mathematical induction method to derive the probability of winning the game as a function of the initial sum of cash n and the target fortune N.

(b) For $p \neq q$, it can be proved that

$$s_n = \frac{1 - r^n}{1 - r^N}.$$

Use the recurrence relation (15.6) to verify this.[2]

Problem formulation.

Part 1. Make a CAS function **one_game(p, n, N)** that takes an integer n (initial cash amount), integer N (target fortune), and a real number $p \in (0, 1)$ (the probability of a win in one round). The function should return a plot of a trajectory S_k of one completed game and a pair of integers, $(k, 1)$ if $S_k = N$ or $(k, 0)$ if $S_k = 0$. In other words, this function will simulate the gambler's rounds of the game until bankruptcy or winning, whichever comes first. Run your function with specific parameters of your choice. Your plot will be similar to the one in Fig. 15.3.

Part 2. Make a CAS function **win_freq(p, n, N, M)** that takes parameters p, n, N of the RW and integer M, runs the function **one_game(p, n, N)** M times, and returns the frequency of winning the game. Run the function **win_freq(p, n, N, M)** with parameters of your choice.

Part 3. Use the frequency of winning found in Part 2 as approximation of the probability of winning. Find the relative error of this approximation.

Part 4. (optional) It is known that the expected number of steps until stopping is given by the formula

$$M(p, n, N) = \begin{cases} n(N - n) & \text{if } r = 1 \\ \frac{1+r}{1-r}\left(n - N \cdot \frac{1-r^n}{1-r^N}\right) & \text{if } r \neq 1 \end{cases}.$$

Design and execute a simulation to compare the theoretical value given by one of these formulas with your numerical estimate.

15.3 Random walk on the square lattice

Consider a simple, symmetric RW on the square lattice $\mathbb{Z}^2$. A walker starts at the origin and at each integer time moves to one of the nearest four lattice points with probability $1/4$ for each direction. Let $S_n = \sum_{j=1}^{n} \mathbf{Z}_j$ be the position of the walker after n steps. Here Z_j is a random vector $[X_j, Y_j]$ taking one of the values $[-1\,0], [1\,0], [0\,-1], [0\,1]$, each with probability $1/4$.

More generally, a popular 2D random walk model is that of a RW on a regular square lattice, but with sites not defined by the integer grid. In this case, at each moment the walker makes a step of the same length a to one of the four neighboring sites.

---

[2]You only need simple algebra to do this.

A constructive rule for choosing the direction of motion can be specified in many ways. The following is one such specification. Generate a random integer r from the discrete uniform distribution with sample space $\{1, 2, 3, 4\}$ and define the direction of the walker's move by the following table:

r_j	Z_j	direction
1	$[-a, 0]$	$\leftarrow$
2	$[a, 0]$	$\rightarrow$
3	$[0, -a]$	$\downarrow$
4	$[0, a]$	$\uparrow$

Alternatively, we can toss two coins and assign one of the four directions of the walker's move to each of the outcomes TT, TH, HT, and HH.

Similar to the one-dimensional RW on integers, one can show (**show this!**) that

$$E(|S_n|^2) = E\left(\left(\sum_{j=1}^{n} X_j\right)^2 + \left(\sum_{j=1}^{n} Y_j\right)^2\right) = n.$$

💡 *Check Your Understanding.* Toss two coins five times and manually plot the corresponding 2D random walk. Use the following outcomes to assign directions:

TT : left; TH : right; HT : down; HH : up.

📖 The concept of a simple random walk can be generalized to higher dimensions. It is known that the formula for the mean square distance from the origin holds for simple, symmetric RWs in any finite dimension.

15.4 Lab 20: Drunken sailor problem

Problem description. Imagine that a drunken sailor comes out of a bar and proceeds in a haphazard way as follows. He makes steps that are always the same length a and moves with equal likelihood forward, backward, left, or right. The level of intoxication is such that the drunken person makes a total of n steps and then lies down and goes to sleep. Fig 15.4 illustrates this.

Now imagine that many drunken sailors leave the bar and move according to the scenario described above. The question that you have to answer using a simulation is: What is the average distance of a drunken sailor from the bar after n steps?

Problem formulation.

Part 1. Give a formal description of the problem in terms of a simple, symmetric 2D RW. Assume that the walker is initially at the origin.

Part 2. Make a CAS function *rw2_distance*(n, a) that takes an integer n and a float a, simulates n steps of length a of the simple, symmetric 2D RW, and returns a plot of the path and the distance $|S_n|$ of the walker from the origin. Run your function for a particular input. Your plot will be similar to the one in Fig. 15.4 (possibly without caricatures of the walker).

Part 3. Make a CAS function *rw2_mean*(m, n, a) that takes positive integers m, n and a float a, runs the function *rw2_distance*(n, a) m times, and returns the average of the output values of the walker distance from the origin.

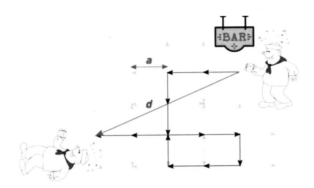

Figure 15.4. A trajectory of a drunken person modeled by a planar two-dimensional simple, symmetric random walk with step a. The length of the red arrow is the distance between the walker and the original position.

Part 4. The theory says that the exact answer for the expectation of the distance of a simple 2D RW is $E(|S_n|) = b_n \cdot a\sqrt{n}$ with certain constants b_n. Run your function *rw2_mean*(m, n, a) with $m = 10000, a = 1$ and $n = 25, 36, 49, 64, 81, 100$. Use your calculations to estimate the constants b_n.

📖 This is not just a fun toy problem. The basic "drunken sailor" random walk problem is widely used in many branches of science to answer important questions regarding natural phenomena. In particular, the sailor walking model corresponds to the motion of an atom migrating on a (square) lattice in 2D. Finding the average distance of a walker from the initial position for a given number of steps has a direct correspondence to the definition of an important physical parameter called the **diffusion constant**.

15.5 Glossary

- **One-dimensional simple random walk on integers** (1D RW) – A sequence of RVs

$$S_n = \sum_{j=1}^{n} X_j, S_0 = 0,$$

where the X_j's are independent RVs with probability space $\{-1, 1\}$ each having a Bernoulli distribution. If the Bernoulli distribution is unbiased, the 1D RW is called **symmetric**. A 1D RW is also called a **random walk on an integer lattice**.

- **Square lattice** – The two-dimensional version of an integer lattice in two-dimensional Euclidean space, denoted as $\mathbb{Z}^2$. It is a regular grid with sites at all points with integer coordinates.

- **Two-dimensional simple random walk on the square lattice** (2D RW) – A sequence of RVs

$$S_n = \sum_{j=1}^{n} [X_j, Y_j], \ S_0 = 0,$$

where X_j, Y_j are independent RVs with the probability space $\{-1, 1\}$ each having a Bernoulli distribution. If the Bernoulli distribution is unbiased, the 2D RW is called **symmetric**.

- **Trajectory** – The result of performing a simulation of a RW (possible path of a random walker).

- **Mean squared displacement** of a RW with N steps – Defined as $E(S_N^2)$. It is known that for a simple symmetric RW in any dimension it is equal to the number of steps N.

- **Absorbing state** (or **absorbing barrier**) – A site on a one-dimensional or two-dimensional lattice such that if the walker reaches this site, the random walk ends.

Appendix A

Data for Lab 17 in Chapter 14

The data for life expectancy at birth for 202 countries is obtained from
    https://www.worldometers.info/demographics/life-expectancy/
It is claimed that the data is based on the latest United Nations Population Division estimates for 2022.

The name of countries are omitted. Each data value gives prediction of average life expectancy at birth for both sexes in a certain country.

$P = [85.29, 85.03, 84.68, 84.25, 84.07, 84.01, 83.99, 83.94, 83.60, 83.52, 83.50,$
$83.49, 83.33, 83.13, 83.13, 83.06, 82.96, 82.94, 82.81, 82.80, 82.80, 82.79, 82.78,$
$82.74, 82.65, 82.48, 82.17, 82.05, 81.88, 81.85, 81.77, 81.55, 81.51, 81.40, 81.17,$
$81.04, 80.94, 80.74, 80.74, 80.73, 80.69, 80.53, 79.89, 79.85, 79.85, 79.64, 79.41,$
$79.27, 79.27, 79.18, 79.18, 79.11, 79.10, 79.02, 78.96, 78.58, 78.46, 78.45, 78.43,$
$78.23, 78.16, 78.00, 77.93, 77.87, 77.74, 77.73, 77.71, 77.56, 77.50, 77.47, 77.47,$
$77.44, 77.43, 77.39, 77.36, 77.33, 77.31, 77.17, 76.79, 76.67, 76.65, 76.57, 76.50,$
$76.47, 76.41, 76.35, 76.26, 76.06, 75.87, 75.85, 75.77, 75.73, 75.69, 75.55, 75.51,$
$75.49, 75.41, 75.23, 75.20, 75.09, 75.05, 75.01, 74.88, 74.65, 74.62, 74.59, 74.28,$
$74.24, 74.08, 74.06, 73.91, 73.90, 73.75, 73.74, 73.58, 73.57, 73.44, 73.38, 73.33,$
$72.99, 72.98, 72.89, 72.77, 72.59, 72.54, 72.50, 72.35, 72.34, 72.32, 72.30, 72.13,$
$72.04, 71.95, 71.76, 71.74, 71.66, 71.32, 71.08, 71.08, 71.01, 70.99, 70.54, 70.53,$
$70.42, 70.26, 70.18, 70.00, 69.86, 69.17, 68.89, 68.87, 68.63, 68.27, 68.21, 67.91,$
$67.87, 67.81, 67.79, 67.78, 67.48, 67.47, 67.03, 66.44, 66.39, 66.09, 65.98, 65.62,$
$65.57, 65.22, 65.21, 65.03, 65.00, 64.99, 64.94, 64.88, 64.86, 64.70, 64.38, 63.62,$
$63.26, 62.98, 62.84, 62.71, 62.64, 62.22, 62.16, 62.13, 62.13, 61.60, 61.05, 60.54,$
$60.32, 59.82, 59.38, 58.75, 58.74, 58.34, 55.92, 55.75, 55.65, 55.17, 54.36]$

Bibliography

[1] Paul R. Halmos, *I want to be a mathematician: An automathography in three parts*, MAA Spectrum, Mathematical Association of America, Washington, DC, 1988, DOI 10.1007/978-1-4612-1084-9. MR961440

[2] Garrett Birkhoff and Saunders Mac Lane, *A survey of modern algebra*, Macmillan Co., New York, N. Y., 1953. Rev. ed. MR0054551

[3] Albert Boggess and Francis J. Narcowich, *A first course in wavelets with Fourier analysis*, 2nd ed., John Wiley & Sons, Inc., Hoboken, NJ, 2009. MR2742531

[4] Gilles Brassard and Paul Bratley, *Fundamentals of algorithmics*, Prentice Hall, Inc., Englewood Cliffs, NJ, 1996. MR1345611

[5] David M. Bressoud, *Factorization and primality testing*, Undergraduate Texts in Mathematics, Springer-Verlag, New York, 1989, DOI 10.1007/978-1-4612-4544-5. MR1016812

[6] Ole Christensen and Khadija L. Christensen, *Approximation theory: From Taylor polynomials to wavelets*, Applied and Numerical Harmonic Analysis, Birkhäuser Boston, Inc., Boston, MA, 2004. MR2046440

[7] Drew Conway, *The data science venn diagram.*, 2010, The Data Science Venn Diagram is Creative Commons licensed as Attribution-NonCommercial. ©Drew Conway Data Consulting, LLC. 2015, used under Creative Commons Attribution-NonCommercial Unported License (https://creativecommons.org/licenses/by-nc/3.0/legalcode)

[8] Richard H. Enns and George C. McGuire, *Computer algebra recipes: An introductory guide to the mathematical models of science; With 1 CD-ROM (Windows, Macintosh and UNIX)*, Springer, New York, 2006. MR2207470

[9] Richard H. Enns and George C. McGuire, *An advanced guide to scientific modeling*, Springer, New York, 2007.

[10] A.N. Kolmogorov et al., *Mathematics: Its content, methods, and meaning*, M.I.T. Press, Cambridge, 1999.

[11] D.E. Knuth et al., *Concrete mathematics: A foundation for computer science*, Addison-Wesley Professional, Upper Saddle River, 1994.

[12] P. Zimmerman et al., *Computational mathematics with sagemath*, 2018.

[13] R. Courant et al., *What is mathematics? An elementary approach to ideas and methods*, Oxford University Press, Oxford, 1996.

[14] William L. Briggs et al., *Calculus: Early transcendentals*, Pearson, Upper Saddle River, 2014.

[15] Dmitry Fuchs and Serge Tabachnikov, *Mathematical omnibus: Thirty lectures on classic mathematics*, American Mathematical Society, Providence, RI, 2007, DOI 10.1090/mbk/046. MR2350979

[16] André Heck, *Introduction to Maple*, 3rd ed., Springer-Verlag, New York, 2003, DOI 10.1007/978-1-4613-0023-6. MR1976957

[17] C. M. Hoffmann, *Geometric and solid modeling: An introduction*, Morgan Kaufmann Pub, Burlington, Massachusetts, 1989.

[18] Gregory F. Lawler and Lester N. Coyle, *Lectures on contemporary probability*, Student Mathematical Library, vol. 2, American Mathematical Society, Providence, RI; Institute for Advanced Study (IAS), Princeton, NJ, 1999, DOI 10.1090/stml/002. MR1707283

[19] Magister_Mathematicae, *A scalable image of conic sections.*, 2012, Wikimedia Commons, File is licensed under the Creative Commons Attribution-Share Alike 3.0 Unported license (https://creativecommons.org/licenses/by-sa/3.0/deed.en).

[20] Victor V. Prasolov, *Polynomials*, Algorithms and Computation in Mathematics, vol. 11, Springer-Verlag, Berlin, 2010. Translated from the 2001 Russian second edition by Dimitry Leites; Paperback edition [of MR2082772]. MR2683151

[21] Hans Rademacher and Otto Toeplitz, *The enjoyment of mathematics; Selections from mathematics for the amateur*, Princeton University Press, Princeton, N. J., 1957. MR0081844

[22] Jan von Plato, *Creating modern probability: Its mathematics, physics and philosophy in historical perspective*, Cambridge Studies in Probability, Induction, and Decision Theory, Cambridge University Press, Cambridge, 1994, DOI 10.1017/CBO9780511609107. MR1265718

Index

1D random walk with boundaries, 226
1D simple random walk (RW), 225
2D simple RW on the square lattice, 225

algorithm, 8
amplitude, 162
argument of a complex number, 28
array data structure, 7

bifurcation diagram, 141
big $\mathcal{O}$ notation, 15
binary number system, 51
bisection method, 107

Central Limit Theorem (CLT), 206
centroid of a triangle, 30
circumcenter of a triangle, 31
cobweb plot, 131
column vector, 22
complex conjugate, 27
computer algebra system, 3
conditional statements, 8
coprime integers, 47
coprime set, 52
Cramer's Rule, 81
cumulative distribution function (CDF), 201

data compression, 187
data structures, 7
depressed cubic polynomial, 74
dictionary data structure, 6
difference quotient, 100
differential of function at a point, 101
Diophantine equation, 47
direction vector, 24
discrete probability space, 195
discrete random variable (RV), 195
discrete uniform distribution, 197
discriminant of a conic section, 40
dot product, 22

eccentric anomaly, 138
eccentricity, 44
envelope of a family of curves in the plane, 70
Euler line, 35
event, 195

expectation (expected value, mean), 196

feasible set of ILP, 62
Fermat pseudoprimes to base a, 54
Fermat's factorization method, 54
Fermat's Last Theorem, 47
Fibonacci sequence, 105
filtering, denoising, 187
frequency, 161
Frobenius number, 62
Fundamental Theorem of Arithmetic, 52
Fundamental Theorem of Calculus (FTC), 115

Gaussian elimination technique, 79
general solution of the 2D LDE, 60
geometric median, 104
Gibbs phenomenon, 166
Gini index, 121
globally nonnegative polynomial, 76
group, 26

hyperbolic cosine, 41
hyperbolic sine, 41

identifier, 7
imaginary unit, 27
inconsistent system of equations, 81
indeterminate system of equations, 81
initial value problem (IVP), 124
integer linear programming problem (ILP), 62
interpolation, 143
inverse matrix, 25
iteration method, 130

Lagrange polynomial interpolants, 147
Law of Large Numbers (LLN), 204
Least Squares, LS, 144
line of best fit, 153
line of perfect equality, 120
linear congruential generators, 210
linear Diophantine equation (LDE), 48
logarithmic fit, 156
loop, 9
Lorenz curve, 120

matrix, 19

mean (average) value of a function over an
    interval, 115
mean squared displacement, 227
modulo function, 13
modulus of a complex number, 28
monic polynomial, 69
monotonic increasing/decreasing sequence, 105
Monte Carlo (MC) integration, 216
multiplicity of polynomial root, 66

normal curve, 203
normal, or normal vector, 23

ordinary differential equation (ODE), 123
orthocenter of a triangle, 31

parallel translation, 25
parametric equations of line, 24
percentile, 204
periodic function, 161
polar coordinates of a point, 28
polar form of a complex number, 28
polynomial equation, 65
polynomial long division, 67
prime number, 52
Prime Number Theorem, 49
prime-counting function, 49
primitive Pythagorean triple (PPT), 47
probability density function (PDF), 201
probability mass function (PMF), 196
probability table, 196
pseudocode, 10
Pythagorean triple (PT), 47

quantile function, 204

rational curve, 80
recursive function, 14
reducible polynomials, 85
reflection transformation, 25
remainder term of Taylor polynomial, 145
residuals, 153
resonance, 163
rigid transformations, 25
roots of polynomial, 65
roots of unity, 66, 73
rotation matrix, 25
row vector, 22
Runge's phenomenon, 150

sample space, 195
scalar projection, 23
semi-major axis, 41
semi-minor axis, 41
sequence, 4
set, 4
shear transformation, 119
Simson line, 37
smooth curve, 101
solution to IVP, 124

spline, spline degree, cubic spline, 150
square lattice, 225
square wave, 159
standard deviation, 196
standardization of a random variable, 206
Stirling's formula, 230
Sylvester matrix, 79, 88

Taylor polynomial, 145
transposition, 21
trial division, 53
trigonometric polynomial, 159

unit vector, 23

variance, 196
Vieta's formulas, 69